# Physics
## —— in the ——
# 21st Century

Proceedings of the
11th Nishinomiya–Yukawa Memorial Symposium

# Physics
## —— in the ——
# 21st Century

Nishinomiya, Hyogo, Japan    7–8 November 1996

Editors

## K. Kikkawa, H. Kunitomo & H. Ohtsubo
Osaka University

**World Scientific**
*Singapore • New Jersey • London • Hong Kong*

*Published by*

World Scientific Publishing Co. Pte. Ltd.

P O Box 128, Farrer Road, Singapore 912805

*USA office:* Suite 1B, 1060 Main Street, River Edge, NJ 07661

*UK office:* 57 Shelton Street, Covent Garden, London WC2H 9HE

**PHYSICS IN THE 21ST CENTURY**

ISBN 981-02-3088-5

Printed in Singapore.

# Contents

The talks "Discovery in the Land of Atoms" presented by D. M. Eigler and "Accelerator
Technology for the Development of Basic Science" by H. Sugawara are not included in the
proceedings, because the manuscripts were not available for publication.

# Preface

The series of the Nishinomiya-Yukawa Memorial Symposium was initiated in 1985 to celebrate the 50th anniversary of the late Dr. Hideki Yukawa's publishment of his meson theory, at the time he had been living in Nishinomiya city. Since then, we have owed the city most of financial support and running administrations. In 1995, on the occasion of the 10th symposium to celebrate the 60th year of the meson theory, we planned a special program which would review the topics of many exciting research areas and provide prospects over the physics in the 21st century. However, due to the tragic earthquake which attacked the city in January 16, 1995, however, we had to abandon to open the symposium. Last year, to our great pleasure, we could again continue the series of symposium supported by the citizen's enthusiaum over the academic acitivities. We wish to express our sincere thanks to Nishinomiya city and the citizens.

The purpose of the present 11th symposium was to realize the previous plan that we invite togeher world leading physicists in various exciting research fields and ask to present reviews over recent acitivities and to provide us prospects over physics in the 21st century. The symposium was in great success. It is our hearty pleasure to share the excitations with readers of the proceedings.

The organizing committee of the symposium consited of

S. Ikeuchi (Osaka University)
K. Kikkawa (Osaka University)
Y. Kuramoto (kyoto University)
Y. Nagaoka (YITP, Kyoto University)
M. Ninomiya (YITP, Kyoto University)
H. Ohtsubo ( Osaka University)
H. Sato (Kyoto University)

The symposium was organized under the auspices of the Education Board of Nishinomiya City and Yukawa Institute for Theoretical Physics, Kyoto University.

Keiji Kikkawa
Hiroshi Kunitomo
Hisao Ohtsubo

May 1, 1997

# Opening Address

*November 7, 1996*

## Junzo Baba
*Mayor of Nishinomiya City*

I am very pleased to open the 11th Nishinomiya-Yukawa Memorial Activities.

This symposium has been held every year since 1986. Unfortunately, however, we couldn't have this symposium last year because of the tragic Hanshin Earthquake.

In Nishinomiya more than 1000 people were killed, and many buildings and other important properties were damaged. At that time, not only the scientists who have joined the symposium, but also the people who have been concerned with Yukawa Memorial Activities in Japan and other countries, encouraged us very much by their letters and generous contributions. We are very grateful to that.

Although the scars of the earthquake are still seen, we can hear the sound of the reconstruction around Shukukawa and others in Nishinomiya. We will make our further efforts to rebuild the city.

As the Mayor of Nishinomiya I would like to express my sincere gratitude to Professor Kikkawa, chairman of the Steering Committee, and the other members of committee for their great efforts payed to achieve the scientific acitivities.

In 1985 many scientists who studied under the guidance of Professor Hideki Yukawa constructed the monument at Kurakuen in Nishinomiya commemorating the birth of the Yukawa meson theory, for which Professor Yukawa was awarded the Novel Prize. Recalling this, we decided to hold the Nishinomiya-Yukawa Memorial Acitivities in order to honor Professor Yukawa's great achievement. We want to take this opportunity for Nishinomiya to be recognized as the birth city of the Meson Theory inside and outside Japan, and to improve citizens' interest in science.

This year, Nishinomiya City will hold the 11th symposium and the memorial lecture will be given at Frente Hall by Professor Hirotaka Sugawara, National Laboratory of High Energy Physics. This time we will award the Nishinomiya-Yukawa Memorial Prize to two young brilliant researchers.

We, citizens will be happy if this project is helpful to the researchers of physics, and if it contributes to the development of fundamental physics.

Lastly I would like to express my sincere thanks for your participation in these activities. Thank you very much.

# Three Stages, Three Modes, and Beyond

Yoichiro Nambu

*The Enrico Fermi Institute*

*University of Chicago*

### Abstract

I review the developement of particle physics from the viewpoint of theoretical methodology. Particle physics has gone through certain characteristic periods in the course of discoveries and accumulation of knowledge, both factual and theoretical. To anlyze this situation, I draw on the theses of three stages and three modes, and illustrate the points with historical examples, up to the present Standard Model and the speculative theories of unification. As an exercise in this analysis, I propose a possible regularity in the quark masses.

## 1  INTRODUCTION

By nature, scientists are optimists. They are forward-looking. I have been told that the main theme of the conference was not to look back, but to look forward and try to anticipate the physics of the twenty-first century. But I do not think I am up to such an ambitious task. Prediction is always difficult, especially prediction of the future, as Niels Bohr was fond of quoting. So I will first look back 50 years of particle physics. They coincide with the post-war years which also more or less covers the bulk of the most important developments of what we call now high energy or particle physics. If my tone sounds rather philosophical, it may be as much due to my age as it is due to the education I received in my early years.

Indeed the title of my talk reveals my age and my background. When I became a physics student at the University of Tokyo some 55 years ago, the Bohr atom was 28 years old, and quantum mechanics was only 15 years old. Compare this with the present day. The main achievement of particle physics is encoded in the Standard Model, which is almost 30 years old. Then the supersymmetry is 23 years old, and the superstring 15 years. Do the youngest generation of theoretical physicists look upon the Standard Model as I had looked upon the Bohr theory? Do they look upon the superstring theory as I had looked upon quantum mechanics? I do not know, but I would not be too surprised if some of them do. On the other hand, even they would admit that, unlike quantum mechanics, supersymmetry and superstring are still speculative

1

theories. I must say that the time scale of progress has lengthened compared to those truly revolutionary days.

## 2 THE THREE STAGES

In my view [1], particle physics owes its origin to Lawrence and Yukawa. With his cyclotron of 1930 Lawrence supplied the basic experimental tool, and with the meson theory of 1935 Yukawa supplied one of the basic theoretical tools. They are very much alive today, and are all taken for granted. In the following I will elaborate on what I mean by theoretical tools. Immediately after the war, my generation of young theorists were strongly influenced by the thinkings of the very original Japanese school of particle physicists, which owes its origin to Y. Nishina, a theorist turned experimentalist. He laid the foundation of nuclear and cosmic ray physics at his Riken (Institute for Physical and Chemical Research) laboratories, which continues to be one of the top research establishments in Japan to this day. He nurtured theorists such as Tomonaga, Yukawa and Sakata, with whom he kept in close contact. Before the war, in 1935, Yukawa initiated the meson theory of nuclear forces. During the war years, Tomonaga, who himself was at Riken, developed the idea of renormalization. Then, with the advancement of experimental techniques, came the discovery of more and more unexpected particles and phenomena: the strange particles, a whole bunch of hadrons, resonances, and parity and CP violation.[2,3]

During these exciting times, S. Sakata and M. Taketani, collaborators of Yukawa in developing the meson theory, articulated their conscious theoretical approach and strategy to particle physics, and practiced it themselves with considerable success.[4]

The Sakata-Taketani methodology is summarized in Taketani's three stage thesis. It posits that the progress in physics, especially particle physics, goes through a cycle of three stages.

### Stage 0

We start with the discovery of new phenomena which are outside of the existing physical laws.

### Stage 1: Phenomenology

Our first task would be to collect data, and try to find some order or regularities in them, then arrive at some empirical laws or representations that enable us to organize them and make predictions. This is the stage of phenomenology.

Examples. The discovery of strange particles and hadron resonances fits this description. The Gell-Mann-Nakano-Nishijima law [5,6] was found by assigning isospin and strangeness to strange particles. This led to flavor $SU(3)$ symmetry of hadrons [7,8,9,10] from which were derived various symmetry relations among masses and cross sections. The Veneziano model [11], which is an explicit formula realizing duality and straight Regge trajectories, might be called another example of this stage.

### Stage 2: Model building

After the first stage of phenomenology comes the stage of model building. We want to interpret the origin of the regularities in terms of a model, in which concrete, often hypothetical, entities are introduced. Or else the model may be mathematical in nature.

The Sakata triplet (proton, neutron, and $\Lambda$) [12] and the successful quark model [13,14] are such examples of realizing the $SU(3)$ symmetry relations in terms of material particles. For the Veneziano formula, the dual string model is its realization in terms of concrete entities called strings.

These models, however, may have large numbers of arbitrary parameters, and even so their validity is of a limited and semi-quantative nature. For different kinds of phenomena there may be different models, and they may not necessarily be mutually compatible.

### Stage 3: Definitive theory

The next and the final stage is to construct or invent a theory which incorporates the various models in a precise, all-inclusive mathematical system of laws, which describes all phenomena in a quantitatively correct way, and can make precise predictions. We immediately recall the Standard Model (SM) as an example, which consitsts of quantum chromodynamics of string interactions and the Salam-Weinberg theory of electroweak interactions. In the past 15 years or so, its validity has been confirmed to an impressive degree of precision. It has all the features of a definitive theory of the three basic interactions. It is still called a model, perhaps because some aspects of it has not yet been confirmed, and perhaps also because of the large numbers of phenomenologically adjustable parameters which could be derived in a more complete theory. Nevertheless its confirmed validity is such that essentially it deserves to be called the Standard Theory.

**Stage 4. Return to Stage 0**

The three stages are expected to repeat themselves, when the final theory is found to break down in the face of new phenomena, which in particle physics usually happens when the available accelerator energy is raised, or the experimental precision is greatly improved. We have been so accustomed to this that we have acquired a habit always expecting it, so that as soon as a theory is established, we immediately start looking for its violations. Here is what Salam had said on this point:

"Classical physical theories are profound. Take the second law of thermodynamics, for instance: Heat cannot flow spontaneously from a colder to a hotter body. Compare this to what you have been doing. You propose some symmetry, and ten seconds later you are already trying to figure out how to break it."

—A. Salam [15]

## 3 THE THREE MODES

No doubt such characterization of particle physics has been most appropriate up to the 1960s. It was the period, as had been in the past, in which experiment and theory went hand in hand to explore and understand nature. In fact, the above analysis could apply equally well to the history of quantum theory. As Sakata had emphasized, the moral of it for the particle physicist is to have in one's mind at any moment a larger picture of the situation, always asking oneself in what stage one is in at the present point in time, and what is likely to follow.

But it is inevitable that even this useful general thesis about the stages of physics itself has its own limits. Indeed, starting around the seventies, there has been a sea change, so to speak, in the way particle physics has evolved. As we have successfully developed and synthesized all the theoretical ideas and experimental facts into the Standard Model, the theory further took off on its own in an exponential way, leaving experimental capabilities far behind. Here the term 'exponential' is a literal one. On the one hand, the theoretical leap is caused by the fact that physical "constants" vary with energy scales logarithmically due to renormalization, so that it becomes natural or even neccesary to envisage new phenomena to appear at exponentially high energies, whereas, on the other hand, the experimental capabilites to test such theories are limited by the worldly realities. It is true that, empirically, the accelerator energy has also grown since the 1930s exponentially with time, by a factor of 10 every 10 years, i.e., a factor of $10^6$ in the past 60 years (the Livingston law [16]). But the GUT and Planck energies that theorists envisage are a factor of $10^{12}$ or more ahead. Without some breakthrough on the part of experimental techniques or methods, the gap looks difficult to close.

In view of this situation, some time ago I proposed [17] a different characterization of the way theories are created and developed. Again it is not a classification of the physical laws themselves but that of the theorists' modes of operation to get at them. As I see it, there are three such modes, and each physicist seems preferentially inclined to one or the other.

## 1. Yukawa mode

This mode can be characterized by saying that one tries to explain or understand new phenomena in terms of concrete entities, i. e., particles. One freely postulates or invents new particles if necessary. Schematically,

$$\text{new physics} \quad \rightarrow \quad \text{new particles}$$

which is to be contrasted with alternative approaches, for example,

$$\text{new physics} \quad \rightarrow \quad \text{new principles}$$

Let us illustrate the point with concrete historical examples:

*Example 1.   Beta decay*

In order to solve the continuum spectra of beta decay, Pauli postulated the neutrino. An alternative which was considered was giving up of the energy conservation hypothesis.[18]

*Example 2.   Nuclear forces*

In order to explain the nuclear forces, Yukawa applied the then new quantum field theory, and postulated the existence of a new particle, now called pion.[19] In order to appreciate its significance, one must recall the prevailing attitude of the leading physicists at that time. They had just succeeded in explaining atomic physics by means of a new physical principle called quantum mechanics. Nuclear phenomena involved many orders of magnitude higher energies and shorter distances. It was not unnatural for the founders of quantum mechanics to expect that to each new physics there corresponded a new mechanics or principle. On the other hand, they had taken for granted that matter consisted of the electron and the proton (perhaps also the neutron), and were reluctant to conceive of new particles just for the sake of understanding the nucleus.

6

*Example 3. The cosmic ray muon puzzle*

The muon discovered in the cosmic rays by Anderson and Neddermyer was at first identified erroneously but understandably with the pion which Yukawa had predicted two years before. But soon theoretical inconsistencies surfaced in the disparity between its production and interaction cross sections. The solution to this puzzle was the two-meson hypothesis[20] of Sakata who assumed that the cosmic ray 'meson' (fermion) was the daughter of the Yukawa meson (boson). On the other hand, a host of theorists tried to explain this away, unsuccessfully, in terms of the strong coupling theory of the Yukawa meson[21]. The situation is akin to example 2 above.(Incidentally, the strong coupling theory has features similar to the Skyrmion theory of the nucleon.)

*Example 4. Two neutrinos and the charm quark*

With the discovery of the muon, it would have seemed natural, in hindsight, to suppose the existence of its own neutrino, but in fact this was not natural or obvious in those days. When the Cabbibo theory was proposed, it would have been natural to expect a fourth quark (charm), but still this did not seem to come so naturally. I will not go into rather complicated historical details, but I would like to emphasize the fact that the Sakata school was among the earliest people to take this path, because, in my view, they did not have psychological resistance to introducing new particles[22].

*Example 5. The three generations*

Continuing in the same direction, it is not surprising that Kobayashi and Maskawa [23] of the Sakata school were led to anticipate the existence of three (or more) generations of fermion, in order to solve the problem of CP violation.

## 2. Einstein mode

In what I call the Eisntein mode, one first sets down a general physical principle to describe a physical phenom, formulate a mathematical theory that realizes it, which will then lead to predictions. Thus one may say that the Yukawa mode is inductive, whereas the Einstein mode is deductive. Of course the latter too starts from some known properties of physical phenomena, but thereafter one is guided by the logic of a theory rather than by the facts. The following examples will need no elaboration.

*Example 1. Theory of gravity*

The logic of Einstein's theory of gravity can be summarized by the following chart.

geometrical principle, equivalence principle $\longrightarrow$ equation of gravity
$\longrightarrow$ predictions $\longrightarrow$ (precession of the perihelion of Mercury,
the black hole, the expanding universe, etc.)

*Example 2. The Gauge theory* – Here the chart is

gauge principle $\longrightarrow$ non-Abelian gauge (Yang-Mills) theory
$\longrightarrow$ QCD, the electroweak theory, grand unified theories

## 3. Dirac mode

The Dirac mode is more abstract and speculative than the Einstein mode. Contact with known physical phenomena becomes more indirect. The primary guiding principle appears to be mathematical beauty and elegance, based on the belief (Dirac [24]) that what is mathematically beautiful must also be physically true. Thus

mathematical possibility $\longrightarrow$ physical theory $\longrightarrow$ predictions

*Example 1. Electric-magnetic duality*

Dirac [24] invented the magnetic monopole (as well as the fiber bundle) in order to realize a symmetry between electric and magnetic forces, solely on the basis of esthetic desirabilty. Much later, the non-Abelian gauge theories were found to be natural realization of the idea in the form of topological solitons. Now this has further developed into the exciting recent developments in the new dualtity principle [25].

*Example 2. Supersymmetry and superstring*

Like the monopole, supersymmetry was not directly motivated by known physics, but by the conceived desirability of unifying fermions and bosons. Nevertheless, because of its mathematical elegance and richness, it has grown into a highly developed theoretical system. Superstring theory may also be placed somehow in this category. Although it is descended from the dual string model, it is a grand and speculative leap. Its concrete appeal is the possibility of unifying gravity and everything.

So far none of the above examples of the Dirac mode has been confirmed experimentally. Nevertheless they are where the center of theoretical activity lies at the moment. The reason for it is not difficult to locate: the fast mathematical developments in this mode coupled with the slow progress on the experimental side, that has been characteristic of particle physics since the 1980s.

To summarize the themes developed, particle physics in the past 60 years has gone through various stages, and in each stage physicists resorted to some dominant theoretical methods appropriate to that stage. We are at present at the near final point of having a complete theory of what is known up to the electroweak energy scale. We have a feeling that the Standard Model is the true theory, and it just remains to confirm it in detail. We have also theories of what to expect beyond the electroweak scale, and various pieces of indirect evidence are called upon in this respect, especially in astrophysics and cosmology, but as yet there are no clear and direct experimental signs to indicate that new physics lies ahead. In a sense, particle physics is at a self-sufficient stage now. A set of principles or paradigms may be said to have guided the physicist to achieve this success:

a. Renormalization

b. The Symmetry and the gauge principle

c. Spontaneous symmetry breaking

As for the future, we expect the following additional principles will play crucial roles, although they remain to be just theoretical expectations at the moment.

d. Supersymmetry

e. Superstring

It is also to be remarked that, theoretical speculations about new physics aside, there are some open problems within the known domain of physics that await further exploration, experimentally as well as theoretically. Outside of the Standard Model, the physics of gravity, including quantum gravity, is an outstanding one. Within the Standard Model, they concern basically the Higgs sector. Experimentally, the masses of the neutrinos are not known yet. The Higgs boson is yet to be discovered. Theoretically, they concern the origin of the masses and mass matrices of quarks and leptons, their hierarchical structure, and the nature of the Higgs field. Since my interests have been in the problem of mass, for which the symmetry breaking paradigm, c, is relevant, I will devote the remainder of the talk to discuss this problem.

# 4 REVIEW OF SPONTANEOUS SYMMETRY BREAKING [26]

The concept of spontaneous symmetry breaking (SSB) has been around for a long time without being recognized as such. As for the concept and use of symmetry in physics, P. Curie [27] consciously applied group theoretical considerations to crystal symmetries, and derived certain selection rules for physical phenomena. He showed that for a phenomenon to occur, for example the Wiedemann effect (magnetization caused by twisting a bar), the medium must have symmetries that are consistent with the symmetries of the effect. He did not realize, however, that the converse is not necessarily true: A symmetry can be broken spontaneously to produce an effect. P. Weiss's explanation of ferromagnetism is an example of SSB. Heisenberg gave a microscopic theory of ferromagnetism in terms of quantum mechanics. It is not surprising, then, that he later invoked an analogy to ferromagnetism in his unsuccessful attempt at a unified theory of elementary particles [28].

SSB is most relevant to the problem of fermion masses. A massless fermion has chiral symmetry, but the latter can be broken spontaneously to produce mass. In other words, mass is a dynamical quantity that is subject to theoretical explanation. The prototype of such an effect is the BCS (Bardeen-Cooper-Schrieffer) mechanism for superconductivity. Other examples are superfluidity in $^3$He, the QCD-chiral dynamics of quark mass (the so-called constituent mass) generation, nucleon-nucleon pairing and low lying collective modes in nuclei. The electroweak symmetry breaking in the Salam-Weinberg theory may be another example if the Higgs field can be interpreted as composite in nature. [26]

The SSB in general has the following characteristics.

1. Degeneracy of the ground state of a medium.

2. Existence of the Nambu-Goldstone (NG) modes when the symmetry is continuous and the system is infinite (the thermodynamic limit)

3. Possibility of hierarchical SSB (tumbling) – This means that an SSB can trigger another in a hierarchical way. Examples of this are the chains

   a) crystal formation (continous to discrete space symmetry) $\longrightarrow$ phonon (an NG mode) $\longrightarrow$ superconductivity

    b) QCD chiral SSB $\longrightarrow$ $\pi$ and $\sigma$ mesons (see below) as the source of nuclear forces $\longrightarrow$ formation of nuclei $\longrightarrow$ nuclear pairing and collecive modes

The BCS mechanism, in addition to the above, has the following special properties.

1. Formation of a fermion mass (energy gap)

2. Existence of the collective bosonic modes $\pi$ (NG) and $\sigma$ (Higgs)

3. A mass formula (approximate) for the low energy fermionic and bosonic modes

$$m_\pi : m_f : m_\sigma = 0 : 1 : 2$$

This can be shown to imply a sort of broken supersymmetry among these modes.

4. Possibility of transcribing the dynamics of these modes into an effective Ginzburg-Landau- Higgs (GLH) theory.[29] In this transcription, the Yukawa coupling $f$ and the self-coupling $\lambda$ of the effective Higgs field are given by

$$1/f^2 = C\ln(\Lambda/m_f), \ \lambda = f^2$$

where $C$ and $\Lambda$ are determined by the underlying dynamics.

## 5   THE STANDARD MODEL AND BEYOND: THE MASS HIERARCHIES

    I come now to the main theme of the conference, i. e., to examine the present and gauge the future, as it relates here to the problem of mass. This may further be broken down to a set of three problems, although they are closely dependent on each other; the solution to one of them will influence the other.

a) Hierarchy within the SM and the details of the Cabbibo-Kobayashi-Maskawa (CKM) matrix

b) The nature of the Higgs field in the SM

c) General hierarchy problem in the GUTS and beyond

The first problem addresses what is already known experimentally: the mass spectrum of quarks and leptons. Still unknown are the neutrino masses and their mixing matrix.

In the second problem, the existence of the Higgs field has been only indirectly confirmed in the gauge sector. The Salam-Weinberg theory gives a precise account of the properties of the electroweak interaction. The existence of the Higgs boson or the content of the pure Higgs sector is an open question. Similarly, the role of the Higgs field in generating the fermion masses is theoretically expected but not experimentally confirmed.

The third problem remains a theoretical one at present. However it has arisen naturally in the process of pursuing the SM to its logical conclusion. In spite of its successes, the SM is not theortically complete. They are unsatisfactory in some respects:

1. It does not unify the strong and electroweak sectors.

2. It does not explain the fermion mass structure, including CP violation and the theta vacuum angle. The search for answers to them has led to the GUTS, but at the same time it has created a new set of problems, like proton decay, monopole, and hierarchy.

It is instructive to first look at the entire spectrum of hierarchy in the universe. In terms of length scale, it spans 60 orders of magnitude, from the size of the present universe ($10^{27}cm$), through various intermediate scales, down to the Planck length ($10^{-33}cm$). In the geometrical middle ($\sim 10^{-3}$) is the biological scale. On the small size side, the scale of the SM spans 6 orders for the charged particles, and perhaps 12 orders including the neutrinos. From the SM to the GUTS scale is again 13 orders. The nature is in fact built on layers of discrete structures and hierarchies. At each layer there is an appropriate set of physical laws which is supposed to be derived as an effective theory from the underlying sublayers by "integrating them out". In between two layers is the domain where a scaling law holds in some form or another. Although one seems to take it for granted, it is this property of natural laws that lies under the success of reductionism. Thus one can start from phenomena of a large scale and find physical laws that apply within that domain, then extend it to smaller scales as much as possible. The Hamiltonian structure of mechanics allows this: The heavenly bodies as well as the bodies on the earth obey the same laws in terms of a few degrees of freedom alone, without having to know their internal structure, and just by changing the parameters. This also holds in the quantum world since basically the Hamiltonian structure remains there. The Wilsonian scaling is another manifestation of this same general property of natural laws. Yet it is also true that such an effective theory has a finite range of validity, and will eventually cease to apply at a critical scale. The next step may be a discontinuous leap, but it still must be such that it permits the process of "integrating out", and the underlying layer decouples to a large extent when pulled back to the original layer. The challenge for particle physics then has been to probe the next layers one by one in this way.

In light of the above considerations, one may say that the three categories of problems listed above regarding the hierarchy at the current microscopic scale are all related and aims to address the same task, but differ in the emphasis or strategy that people want to follow. In concrete terms, I see three approaches to the mass problem:

1. Top down appproach

   Here one starts from a grand scheme like the GUTS at a high energy scale, which was originally motivated by an attempt for unification in the gauge sector, then comes down again to the low energy scale and try to see if it can also explain the mass problem. If successful, it will certainly reinforce our confidence in grand unification.

   Currently this seems to be the preferred strategy of theorists. Thus the various GUT theories like $SU(5)$, $SO(10)$ and higher and richer unification groups up to the superstring level have been considered. At the moment the most promising version of gauge unification seem to be the supersymmetric one, and it is also most interesting one in that the supersymmetric partners are expected to show up in the Tev region. The nature of the Higgs fields in the SM is generally regarded as elementary, i. e., they exist already at the gauge unification scale.

## 2. Bottom up approach

This may be said to be repeating anew of the process of arriving at grand unification applied to the mass problem. For this one examines the mass spectrum in the SM and tries to find some clues there without prejudice, and see how much of it can be understood within the SM, and if not, what new physics would be needed.

The top quark condensation model[30,31,32] belongs to this category. It has achieved some success in making clear that the large top quark mass is more natural than the smallness of the other fermion masses. The Higgs boson is considered as a $t - \bar{t}$ composite here, the only input parameter is the cutoff energy, with the top and the Higgs masses satisfying the BCS relation. But it fails to explain the other lighter fermions. The real origin of the $t - \bar{t}$ binding force is left open, and to be sought eventually in a unification theory at the cuoff.

## 3. Intermediate approach, represented by the technicolor models [33,34]

Here a new set of gauge fields specifically responsible for mass generation is postulated at an intermediate scale, not far above the electroweak scale, from which one comes down to account for the masses in the SM. Going up, they will be unified together with the existing forces in the SM.

It seems fair to say that so far none of these approaches has been so successful as to exclude the other possibilities. The difficulty is in giving a natural explanation for the seemingly irregular hierarchical pattern of masses from a naturally regular and simple input. One does not know yet whether the fermion masses are rooted in a relatively simple law or are a result of the whims of the renormalization group equations. In view of this situation, I will present below, as an exercise in the stage 1 of the Sakata scheme, a possible regularity in the quark masses which may have not been noticed before.

The quark masses are not directly observable quantities, but have to be inferred indirectly and are subject to theoretical uncertainties, hence also allow some degree of freedom of choice. Some of them are especially sensitive to the QCD $\Lambda$ and the renormalization point referred to. With this proviso they seem to satisfy the following simple formula

$$m/m_0 = 2^n,$$

where $m_0 \approx 5$ Mev, and n runs over the following set of integers

14

Table 1: The quark mass formula

| I | II | III |
|---|---|---|
| u: 0 | c: 8 | t: 15 |
| 6±2 | | 6±1 |
| d: 1 | s: 5 | b: 10 |

The numbers inserted between the quark entries are the differences between neighboring generations for up/down type quarks This discrete scaling law is ominously reminiscent of the Bode's law for the radii of the planetary orbits, which is not known to have a theoretical basis. This new one may not have a theoretical basis either. But it might serve as a clue in the search for a proper theoretical starting point. For example, it lends itself to the type of hierarchy theories proposed by Frogatt and Nielsen, perhaps with three different kinds of charges assigned to each quark. Or else it suggests some kind of quantization in scaling. [a]

# 6   CONCLUDING REMARKS

Within the last quarter century, the nature of particles physics has undergone radical changes the likes of which have not been seen before. On the theoretical side, the success of the Standard Model means that one has a working theory that enables one to describe all known phenomena in a precise and comprehensible way. But it still has some phenomenological elements; it is not a complete closed theory that is expected to hold much beyond the electroweak energy scale. Currently, the neutrino and Higgs sectors are unknown and untested. At the same time, one has also developed a class of speculative unified theories that have gone way beyond the energies conceivably accessible by traditional experimental techniques yet are able to make some statements that are subject to experimental tests. The superstring theory raises the hope of bringing gravity within the fold of quantum field theory and accomplish the ultimate unification of all known forces. On the experimental side, however, it is clear that one is slowly approaching the practical limits of the traditional accelerator-based physics. The empirical Livingston scaling law for the exponential increase of accelerator energy has served us well for the past 60 years,

---

[a]The leptons do not conform to such a formula as nicely as the quarks, but if one wishes, one could assign the following numbers:  e: -3.5 $\mu$: 4.5 $\tau$: 8.5

but here again a new qualitative break has to occur if one is going to be able to continue our quest to the foreseeable future, and to be able to test the unified theories in its full content.

The first thirty years of particle physics was like the time of Renaissance. Experiment and theory went hand in hand. Since then, however, we have been getting into the baroque period, so to speak, with increasing levels of sophistication in both hardware and software, and yet with a mismatch between the two. This is the way the things are, whether one likes it or not. According to theory, particle physics has a bright future. Supersymmetry may be around the corner. Superstring may be able to make a connection with reality. Still, as a person from the older era, I feel an urge to close this talk with a couple of quotes

"As a mathematical discipline travels far from its empirical source, or still more, if it is second and third generation only indirectly inspired by ideas coming from 'reality', it is beset with very grave dangers. It becomes more and more purely aesthetizing, more and more purely *l'art pour l'art*. This need not be bad, if the field is surrounded by correlated subjects, which still have closer empirical connections, or if the subject is under the influence of men with an exceptionally well-developed taste. But there is a grave danger that the subject will develop along the line of least resistance, that the stream, so far from its source, will separate into a multitude of insignificant baranches, and that the discipline will become a disorganized mass of details and complexities. In other words, at a great distance from its empirical source, or after much' 'abstract' inbreeding, a mathematical subject is in danger of degeneration. At the inception the style is usually classical; when it shows signs of becoming baroque, then the danger signal is up. — In any event, whenever this stage is reached, the only remedy seems to be the rejuvenating return to the source; the reinjection of more or less directly empirical idea. I am convinced that this was a necessary condition to conserve the freshness and the vitality of the subject and that this will remain equally true in the future."

—— John von Neumann [35]

" Eppure si muove"    — Galileo

## ACKNOWLEDGEMENTS

This work was supported in part by the National Science Foundation, Grant #PHY 91-23780, and the University of Chicago. I also thank Prof. K. Kikkawa of Osaka University for his hospitality.

16

1. Y. Nambu, Prog. Theor. Phys. Suppl. 85 (1985) 104.
2. Articles in Prog. Theor. Phys. Suppl. 105 (1991) 105.
3. L. M. Brown, Historia Scientiarum #36 (1989) p.1;
4. See the articles by Sakata and Taketani in Prog. Theor. Phys. Suppl. 50 (1971).
5. M. Gell-Mann, Phys. Rev. 92 (1953) 833.
6. T. Nakano and K. Nishijima, Prog.Theor. Phys. 10 (1953) 581.
7. M. Gell-Mann, CalTech Report CTSL/20 (1964).
8. Y. Ne'eman, Nucl. Phys. 26 (1961) 222.
9. M. Ikeda, S. Ogawa, and Y. Ohnuki, Prog. Theor. Phys. 22 (1959) 715.
10. Y. Yamaguchi, Prog. Theor. Phys. Suppl. N0.11 (1959) 1, 37.
11. G. Veneziano, Nuovo Cimento 57A (1968) 190.
12. S. Sakata, Prog. Theor. Phys. 16 (1956) 686.
13. M. Gell-Mann, Phys. Lett. 8 (1964) 214.
14. G. Zweig, CERN reports 8182/TH 401.
15. Quoted in J. J. Sakurai, Ann. Phys., 11 (1960), 1.
16. M. S. Livingston, High Energy Accelerators (Interscience Publishers, 1958), p. 149.
17. Y. Nambu, Prog. Theor. Phys. Suppl. 85 (1985) 104.
18. See A. Pais, in *Twentieth Century Physics*, eds. L. Brown, A. Pais, and B. Pippard, IOP Publishing Ltd, and AIP Press, Inc., 1995), p. 43
19. H. Yukawa, Proc. Phys.-Math. Soc. Japan 20 (1937) 712.
20. S. Sakata and T. Inoue, Prog. Theor. Phys. 1 (1946), 189.
21. See L. M. Brown, in *Twentieth Century Physics*, eds. L. Brown, A. Pais, and B. Pippard, IOP Publishing Ltd, and AIP Press, Inc., 1995), p. 357.
22. See V. Fitch and J. Rosner, in *Twentieth Century Physics*, eds. L. Brown, A. Pais, and B. Pippard, IOP Publishing Ltd, and AIP Press, Inc., 1995), 635; ?? Z, Maki and T. Nakagawa, Prog. Theor. Phys. 31 (1964) 115; Z. Maki, Prog. Theor. Phys. 31 (1964) 331, 333.
23. M. Kobayashi and T. Maskawa, Prog. Theor. Phys. 49 (1973) 652.
24. P. A. M. Dirac, Proc. Roy. Soc. A133 (1931) 60.
25. N. Seiberg and E. Witten, Nucl. Phys. B426 (1994) 19; B431 (1994) 484.
26. See Y. Nambu, in *Eolutionary Trends in Physical Sciences* (Springer Proceedings in Physics, Vol. 57), eds. M. Suzuki and R. Kubo (Springer-Verlag Berlin Heideberg 1991), p. 51; Proc. ICTP Workshop on Quarks and Leptons, 1996, to be published, (EFI preprint 96-39 ).
27. P. Curie, J. Phys. 3 (1907) 393.
28. H. P. Duerr,W. Heisenberg, H. Mitter, S. Schlieder, and K. Yamazaki, Z. f. Naturforschung 14A (1959) 441.

29. V. L. Ginzburg and L.D. Landau, Zh. Exp. Teor. Fiz. 20 (1940) 1064.
    P. W. Higgs, Phys. Lett. 13 (1964) 132; Phys. Rev. Lett. 13 (1964) 508.
30. Y. Nambu, in *New Theories in Physics, Proc. XI Int. Symposium on Elementary Particle Physics*, Eds.: A. Ajduk et al., (World Scientific, Singapore, 1989)
31. A. Miransky, M. Tanabashi, and K. Yamawaki, Mod. Phys. Lett. A4 (1989) 1043; Phys. Lett. B221 (1989) 177
32. W. A. Bardeen, C. T. Hill and M. Lindner, Phys. Rev. D41 (1990) 1647.
33. S. Weinberg, Phys. Rev. D13 (1976) 974, D19 (1979) 1277.
34. D. Susskind, Phys. Rev. D20 (1979) 2613.
35. John von Neumann, in *The Works of the Mind*, (University of Chicago Press, 1947), p. 196

# QUANTUM TRANSPORT
# IN MESOSCOPIC SEMICONDUCTOR STRUCTURES

Tsuneya ANDO

*Institute for Solid State Physics, University of Tokyo*
*7-22-1 Roppongi, Minato-ku, Tokyo 106, Japan*

A brief review is given of quantum transport phenomena in mesoscopic systems. The topics include ballistic transport such as the conductance quantization in quantum point contacts, magnetic focusing, and various phenomena appearing in quantum wire junctions, universal conductance fluctuations arising from quantum interferences, and single electron tunneling due to the charging energy in quantum dots.

## 1 Introduction

Recent advances in microfabrication and crystal growth technology have enabled the preparation of lateral structures with submicron scale such as quantum wires, dots, point contacts, and antidots on the surface of two-dimensional (2D) electron system with mean free path as large as 100 $\mu$m. The transport in this system is ballistic, i.e., electrons are scattered from the confinement potential itself rather than impurities. The purpose of this paper is to give a brief review on various new phenomena observed in such quantum structures and corresponding new concepts important from the point of view of fundamental physics.

Mesoscopic systems are introduced in Sec. 2. Various topics related to ballistic transport are described in Sec. 3. A Coulomb blockade and single electron tunneling through a quantum dot are discussed in Sec. 4. Section 5 is devoted to a history and a future outlook.

## 2 Mesoscopic Semiconductor Structures

### 2.1 Two Dimensional System

Mesoscopic structures are fabricated using two-dimensional (2D) systems. A typical 2D system is realized at a metal-oxide-semiconductor (MOS) structure as is schematically illustrated in Fig. 1. Electrons are induced at the interface of the insulating oxide and the p-type semiconductor when a large bias voltage is applied between the metallic gate and the semiconductor. These electrons form a 2D system because the motion perpendicular to the interface is quantized into discrete energy levels while the motion parallel to the interface remains free. This 2D system is called an inversion layer.[1] The electron concentration in the inversion layer is controlled almost freely as it is proportional to the applied

20

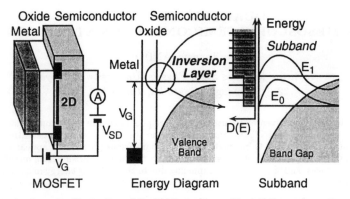

**Fig. 1** A schematic illustration of the MOS structure. The left figure shows the structure, the middle the energy diagram, and the right quantized energy levels and corresponding wave functions for the motion perpendicular to the interface.

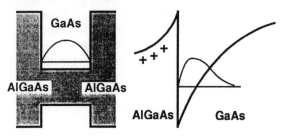

**Fig. 2** A schematic illustration of a GaAs/AlGaAs quantum well (left) and single heterostructure (right).

voltage. The integer quantum Hall effect was discovered in the inversion layer for the first time.[2]

A semiconductor superlattice can be fabricated using the crystal growth technique such as molecular beam epitaxy (MBE) and metal-organic chemical vapor deposition (MOCVD). Figure 2 shows a quantum well consisting of a GaAs layer with width 100 Å – 200 Å sandwiched by AlGaAs layers. The motion perpendicular to the interface is quantized and a 2D electron system is realized using the so-called modulation doping in which only the barrier layer is doped with donors. When the width of the well becomes sufficiently large, we have two independent 2D systems, each of which is localized in the vicinity of the GaAs/AlGaAs interface. This single heterostructure is essentially the same as the MOS structure except that the barrier consists of the semiconductor having a large band gap rather than insulating oxide. The supreme quality of the heterointerface has made this 2D system almost ideal. In fact, the mean free path over which an electron follows a straight trajectory before being scattered

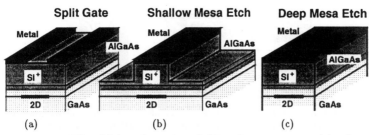

**Fig. 3** Quantum Wires fabricated at a GaAs/AlGaAs heterostructure. (a) split gate. (b) shallow mesa etching. (c) deep mesa etching.

can be as large as 100 $\mu$m. The fractional quantum Hall effect has been observed in this single heterostructure for the first time.[3]

## 2.2 Microfabrication and Lithography

Semiconductor devices are becoming smaller and smaller and the minimum feature size has reached the diffraction limit imposed by the visible light, i.e., a few thousands angstrom. Various new technologies are being developed to go beyond this limit. The technology to fabricate very fine structures has advanced considerably. When we use an electron beam, for example, we can fabricate a wire with a width of the order of a few hundred angstrom. Usually the electron-beam lithography is limited by a large angle scattering in a substrate although the spot size of the electron beam can be reduced much smaller.

Figure 3 shows a schematic illustration of the method of fabricating a quantum wire at a GaAs/AlGaAs heterostructure. In the split-gate configuration (a), 2D electrons are all depleted except in the narrow region below the spacing of the gates under a negative bias on the metallic gates, leading to a quantum wire. A wire can be created also by etching off a part of the AlGaAs layer (shallow mesa etching) or down to the GaAs layer (deep mesa etching). The typical width of the wire is $W > 500$ Å. There are various other ways to fabricate quantum wires.

One important length scale is the Fermi wave length $\lambda_F$ of the 2D system. When a 2D electron is confined into a wire with width $W$, the motion perpendicular to the wire is quantized into discrete energy levels and the states consist of one-dimensional (1D) subbands. For an abrupt confinement potential, the energy levels are given by $E_n = (\hbar^2/2m)(2\pi n/W)^2$ with $n = 1, 2, \ldots$. The number of occupied 1D subbands, i.e. below the Fermi energy $E_F$, is called a channel number. The channel number is obtained as $n = [2W/\lambda_F]$ by using the fact that $E_F = (\hbar^2/2m)(2\pi/\lambda_F)^2$. This leads to $n \gtrsim 2$ for wires with width $W \gtrsim 500$ Å. In the case of metallic wires the Fermi wavelength is of the order of a few angstrom and therefore the channel number is larger than 1000 and

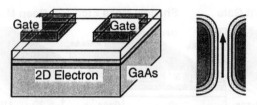

**Fig. 4** A quantum point contact created by a metallic gate with a narrow spacing (left) and a schematic illustration of the equipotential lines for 2D electrons (right).

the quantization into 1D subbands is not important.

Macroscopic systems are characterized by the length of the order of 1 cm or 1 mm. The size of atoms and molecules is between $\sim 1$ Å and $\sim 10$ Å and is called microscopic. The semiconductor structure is characterized by the length of the order of a few thousands angstrom which lies between macroscopic and microscopic scales and is therefore called a mesoscopic system.

The mesoscopic systems provide a new physical system in which electrons are confined by an artificial potential and their electron number or density can be almost freely controlled. They are new quantum many-body systems in which both quantum effects and those of electron-electron interactions are quite strong. A typical example of the manifestation of strong electron-electron interactions can be found in the so-called Tomonaga-Luttinger liquid, the Coulomb blockade in quantum dots, antidot lattices, etc. One fundamental problem on the transport is the crossover from macroscopic to microscopic transport as well as interference effects, corresponding conductance fluctuations, and roles of classical and quantum chaos. The tunneling in mesoscopic systems provides various important problems such as macroscopic quantum tunneling in many electron systems and the tunnel time.

The mesoscopic semiconductor structure has been attracting much attention in the field of electronics. It is closely related to the future of LSI. Various new devices using quantum effects have been proposed and their feasibilities have been studied. A typical quantum phenomenon can be found in the electron wave and its phase. Further, the ballistic conduction without being scattered from impurities and the Coulomb blockade and single-electron tunneling consist another possibilities.

## 3 Ballistic Transport

### 3.1 Conductance Quantization in Quantum Point Contacts

Figure 4 shows a schematic illustration of a quantum point contact with a split gate. Under a strong negative bias voltage on the gate, the 2D electron system is split into two independent systems. When the voltage is appropriate, there appears a narrow conducting path below the splitting of the gate. The

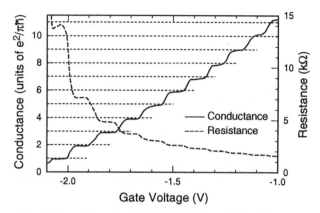

**Fig. 5** An example of experimental results showing the conductance quantization across a quantum point contact.

conductance across the path called a quantum point contact is quantized into $e^2/\pi\hbar$ times an integer.[4,5]

Figure 5 shows an example of experiments. The conductance vanishes when the gate voltage is negative and its absolute value is large. With increasing the gate voltage the conductance exhibits a step-function like increase with the step height being $e^2/\pi\hbar$. Each step corresponds to the opening up of a new conducting channel in the quantum point contact.

### 3.2 Magnetic Focusing

One of the most typical phenomena showing the ballistic nature of the electron motion in a heterostructure is the so-called magnetic focusing. Figure 6 gives an illustration of the experimental configuration. Electrons emitted from the right quantum point contact are collected by the left point contact. In magnetic fields the electron trajectory follows a cyclotron orbit and electrons are focused onto the collector contact at certain magnetic fields (1, 2, and 3 in the figure).

An example of experiments is shown in Fig. 7. The vertical axis shows the voltage proportional to the current going into the left point contact. The peaks 1, 2, and 3 correspond to the orbits 1, 2, and 3 in Fig. 6. Electrons emitted from the right point contact in a direction slightly away from the vertical direction are all focused onto the left point contact when the cyclotron diameter is equal to the distance between the contacts. The same occurs when electrons are reflected by the boundary if it is sufficiently smooth.

### 3.3 Quantum Wire Junctions

Figure 8 shows an example of the configuration in which the transport

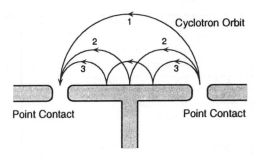

**Fig. 6** A schematic illustration of the magnetic focusing. An electron emitted from the right point contact follows a circular cyclotron trajectory and is collected by the left point contact.

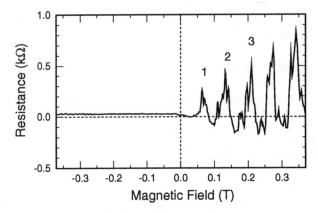

**Fig. 7** Experiments showing the magnetic focusing peaks. The peaks 1, 2, and 3 correspond to the orbits 1, 2, and 3 in Fig. 6.

coefficients of a quantum wire are measured.[6] The conventional resistance is determined by the voltage between terminal 1 and 2 for a current flowing between terminal 3 and 6. The bend and transfer resistance is defined as the voltage between 1 and 3 and between 2 and 3, respectively, for the current between 5 and 6. The transfer resistance vanishes for macroscopic wires. The Hall resistivity is measured from the voltage between 1 and 5 for the current between 3 and 6.

Some examples of measured resistivities are shown in Fig. 9. The bend resistance becomes negative in the absence of a magnetic field and becomes small in magnetic fields. The Hall resistivity exhibits anomalies called quenching and last plateau.[7-11] These properties are the direct result of the ballistic nature of the electron motion in quantum wires.

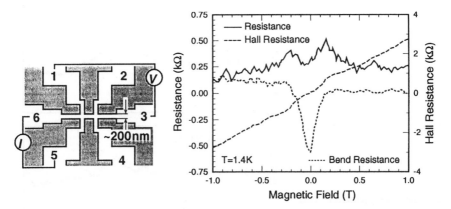

**Fig. 8** (left) Quantum wire junctions for the measurement of transport quantities of quantum wires. The current flows between two of 1~6 terminals and the voltage difference is measured between two of 1~6 terminals.

**Fig. 9** (right) An example of magnetoresistances measured in quantum wire junctions. The solid line represents the conventional resistance (the current flows between 6 and 3 and the voltage is measured between 1 and 2), the dotted line the bend resistance (the current between 6 and 5 and the voltage between 1 and 3), and the dashed line the Hall resistance (the current between 6 and 3 and the voltage between 1 and 5).

## 3.4 Landauer's Formula

The theoretical basis of understanding transport properties of quantum structures is called Landauer's formula,[12] which relates the conductance of a 1D system to the transmission and reflection coefficients. We have

$$G = \frac{e^2}{\pi\hbar}T, \tag{3.1}$$

where $G$ is the conductance and $T$ is the transmission probability at the Fermi level. This formula was given in 1957 when the Kubo formula was proposed for the conductivity in macroscopic systems.[13] There have been various discussions on the relation between Landauer's formula and Kubo formula.[14-16] Landauer's formula can be extended to the case of multi-channel wires by summing up over indexes for in-coming and out-going channels. In multi-terminal cases the corresponding formula is called the Büttiker-Landauer formula.[17]

## 3.5 Series of Two Point Contacts

In the following an example of such theoretical analyses will be discussed. Figure 10 shows a series of two quantum point contacts.[18,19] An electron is emitted from the top contact and collected by the bottom contact. The purpose of this experiment was a direct observation of the wave function of the electron

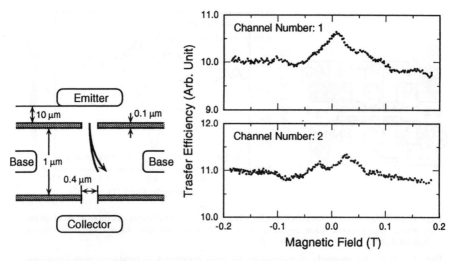

**Fig. 10** (left) A schematic illustration of series of two point contacts.

**Fig. 11** (right) Measured magnetic-field dependence of the conductance across the series of two point contacts shown in Fig. 10. A single peak and double peaks are observed when the channel number of the emitter contact is one and two, respectively.

emitted from the point contact. For this purpose, the electron trajectory is modified by a magnetic field applied perpendicular to the 2D system.

When there is only a single channel in the emitter point contact, the wave function of the emitted electron is expected to have a single peak as a function of the magnetic field. When two channels are present, on the other hand, the absolute value of the wave function is expected to have two peaks away from the vanishing magnetic field. In fact, the experimental results shown in Fig. 11 agree with these expectations.

Some examples of the results of theoretical calculations are shown in Fig. 12.[20] The conductance vanishes in a sufficiently strong field because the electron trajectory curved into the perpendicular direction. The figure (a) shows an expected behavior depending on the channel number of the emitter contact, i.e., a single peak and double peaks when the channel number is one and two, respectively.

The figure (a) corresponds to the conductance at $T \sim 3$ K. On the other hand, a large oscillation appears at lower temperatures as shown in (b). This oscillation is a result of interferences of waves reflected at different boundaries and has an amplitude of the order of $e^2/\pi\hbar$. The classical electron motion in the system is quite complex because of many reflections from the boundary and therefore the interference pattern is also not so simple. This fluctuation of

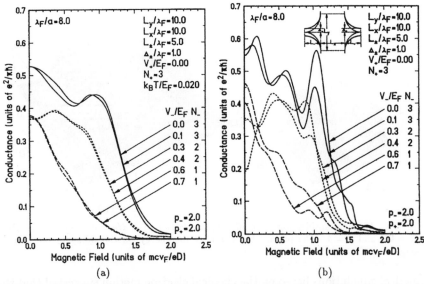

**Fig. 12** Examples of calculated conductance across a series of two point contacts. (a) At temperature of $T \sim 3$ K. (b) At absolute zero of temperature. The dashed lines correspond to the case that the channel number of the emitter contact is one, the dotted lines two, and the solid lines three.

the conductance of the order of $e^2/\pi\hbar$ due to interferences is actually known as universal conductance fluctuations. At higher temperatures like in (b) the interference fluctuations disappear due to the averaging over electron waves having different energies.

### 3.6 Antidot Lattices and Chaos

The classical electron in semiconductor quantum structures is usually chaotic. Antidot lattices are one of such systems that the chaos plays important roles in their transport properties. Figure 13 shows a schematic illustration of a square antidot lattice. The diameter of each antidot is usually about 1000 Å. The antidot lattice is a simplest form of man made lattices although it has a large period $a \gtrsim 2000$ Å. The lattice system is known to comprise with self-similar energy levels called Hofstadter's butterfly in magnetic fields.[21] The antidot lattice can be regarded as a kind of the Sinai billiard in which the chaotic electron motion has been studied intensively.

Figure 15 shows an example of observed diagonal and Hall resistivity in a patterned and unpatterned 2D system.[22] Two prominent peaks appear in the diagonal resistivity $\rho_{xx}$ in weak magnetic fields and correspondingly some step-like structures are present also for the off-diagonal Hall resistivity $\rho_{xy}$. Various

28

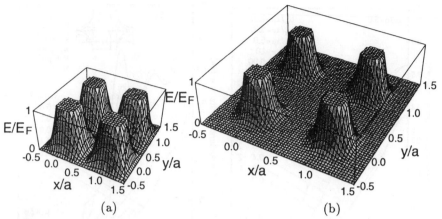

**Fig. 14** A model antidot potential in a square lattice. (a) $d/a = 0.5$ and $\beta = 1$. (b) $d/a = 0.3$ and $\beta = 4$. The potential is cutoff at the energy corresponding to the Fermi energy.

numerical simulations based on the classical electron motion suggested that the chaotic motion of electrons is important as the origin of the commesurability peaks.[23-26]

It is easy to understand the origin of such commensurability peaks in the limit of small aspect ratio $d/a \ll 1$. In this case, the electron loses its previous memory on the direction of the velocity when it collides with an antidot and therefore successive scattering with antidots can be approximately regarded as independent of each other. This means that antidots are nothing but independent scatterers.

Consider the case in high magnetic fields. In high magnetic fields, the transport is possible through the migration of the center of the cyclotron motion and therefore the conductivity vanishes in the absence of scattering. When $2R_c < a$, the scattering of an electron from an antidot cannot give rise to diffusion or conduction because the electron is trapped by the antidot. The situation is certainly different from the case of the random distribution of scatterers of high concentrations in which there is a nonvanishing probability for an electron to be scattered by neighboring another scatterer and the center can migrate due to successive scattering from different impurities.

The scattering from antidots starts to contribute to the conductivity when $2R_c > a - d$. The migration of the center of the cyclotron orbit occurs most frequently due to successive scattering from nearest-neighbor antidots at the magnetic field corresponding to $2R_c = a$. At this magnetic field the measure of such orbits becomes maximum in the phase space due to a kind of the magnetic focusing effect. In fact, a slight change $\theta$ in the direction from the direction

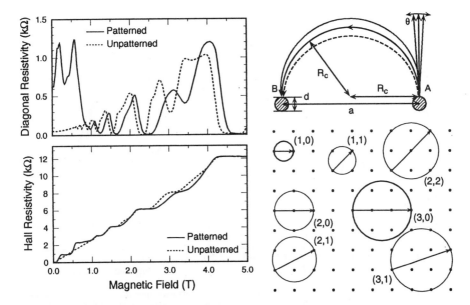

**Fig. 15** (left) An example of $\rho_{xx}$ and $\rho_{xy}$ observed in patterned (solid lines) and unpatterned (dotted lines) 2D systems.[22] Two prominent peaks and step structures appear in the diagonal and Hall resistivity, respectively, in antidot lattices.

**Fig. 16** (right top) A schematic illustration of the magnetic focusing leading to the fundamental commensurability peak.

**Fig. 17** (right bottom) Some examples of cyclotron orbits contributing to the migration of the guiding center. The orbits (1,0) and (3,0) not disturbed by the presence of other dots are expected to make a large contribution to the migration. The other dots are strongly disturbed by other dots and make only a small contribution. It is likely that the orbit (2,0) and (2,1) give a single peak around $2R_c \lesssim \sqrt{5}a$ (the area in the phase space for the latter is expected to be twice as large as that of the former) and further the orbits (2,2), (3,0), (3,1) give a peak around $2R_c \sim 3a$.

normal to the line connecting neighboring antidot leads to a change only of the order of $2R_c\theta^2$ in the position when the electron collides with a neighboring antidot as is shown in Fig. 16. This leads to the increase of the phase-space volume of the runaway and similar orbits contributing to the increase of the diffusion coefficient at $2R_c \approx a$.

As is shown in Fig. 17, the orbit corresponding to $2R_c = a$ can be denoted as $(n_x, n_y) = (\pm 1, 0)$ or $(0, \pm 1)$, where the line segment connecting the point $(n_x, n_y)$ and the origin constitutes the diameter of the circle. With the further decrease in the magnetic field the successive scattering with next nearest-neighbor antidot becomes possible and the conductivity has a peak around

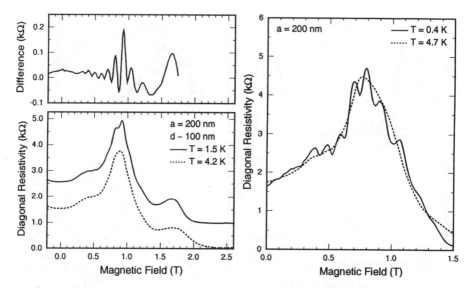

**Fig. 18** (left) An example of observed quantum oscillations superimposed on the fundamental commensurability peak.[27] The oscillation becomes clear in the difference of the resistivity at $T=4.2$ K and that at $T=1.5$ K (shifted vertically by 1 kΩ).

**Fig. 19** (right) An example of observed quantum oscillation superimposed on the fundamental commensurability peak.[28]

$2R_c = \sqrt{2}a$ corresponding to $(n_x, n_y) = (\pm 1, \pm 1)$. This contribution becomes less prominent, however, because the orbit passes through the position of a nearest-neighbor antidot. The next peak arises from $(n_x, n_y) = (\pm 2, 0)$ or $(0, \pm 2)$ and $(n_x, n_y) = (\pm 2, \pm 1)$ or $(\pm 1, \pm 2)$, which lie close to each other. The latter contribution should be larger because its measure is twice as large as that of the former and therefore the peak occurs roughly around $2R_c \sim \sqrt{5}a$. This peak should be weaker, however, as these orbits also pass through the position of other antidots. The next prominent peak is expected to be given by $(n_x, n_y) = (\pm 2, \pm 2)$, $(\pm 3, 0)$, $(0, \pm 3)$, $(\pm 3, \pm 1)$, and $(\pm 1, \pm 3)$. The peak is expected to be at $2R_c \approx 3a$, because the orbits $(n_x, n_y) = (\pm 3, 0)$ and $(0, \pm 3)$ are not disturbed by other antidots. This explains most of the commensurability peaks observed experimentally.

The situation becomes difficult for antidot lattices with a large aspect ratio, because the correlation between the electron motion before and after a collision with an antidot becomes more and more appreciable with the increase of the aspect ratio.[29] Further, the commensurability peaks in the resistivity are affected also by structures appearing in the Hall conductivity because the diagonal conductivity becomes comparable to the Hall conductivity. A large aspect

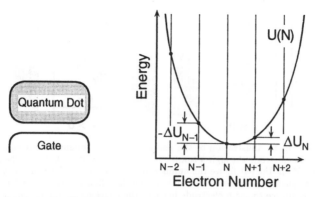

**Fig. 20** A schematic illustration of a quantum dot and its electrostatic energy as a function of the number of electrons trapped in the dot.

ratio corresponds usually to a small antidot period $a$ and to a narrow spacing between the nearest-neighbor antidot comparable to the Fermi wavelength $\lambda_F$. Therefore quantum effects can be important for antidot lattices with large $d/a$. In fact, a quantum mechanical calculation in a self-consistent Born approximation gave the result for $d/a = 0.5$ that the diagonal conductivity $\sigma_{xx}$ exhibits essentially no structure and the off-diagonal Hall conductivity has a small dip at $2R_c \sim a$, leading to the fundamental commensurability peak at $2R_c \sim a$.[30,31]

A fine oscillation was observed superimposed on commensurability peaks of the magnetoresistance.[27-32] Figures 18 and 19 give examples of observed quantum oscillation superimposed on the fundamental commensurability peak reported in Refs. 27 and 28. The period is roughly given by $\Delta B \sim \Phi_0/a^2$ with $\Phi_0 = ch/e$ being the magnetic flux quantum. This oscillation has been analyzed in terms of the periodic orbit theory giving a semiclassical quantization in chaotic system and has been shown to be closely related to quantized energy levels associated with a periodic orbit encircling a quantum dot.[30,33-37]

## 4  Single Electron Tunneling

### 4.1  Quantum Dot and Charging Energy

When electrons are trapped into a finite region like quantum dots, the charging energy or the static Coulomb energy plays a important role. The electrostatic Coulomb energy for a dot containing $N$ electrons has a term proportional to $N(N-1)e^2/2$ on the average corresponding to the number of electron pairs interacting with each other. This can be approximated as $N^2e^2/2$ for a sufficiently large $N$. There are positive charges $+Ne$ somewhere which ensure the charge neutrality condition and the interaction with such positive charges leads

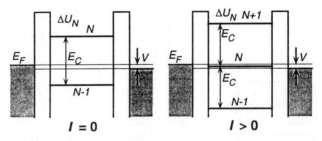

**Fig. 21** A schematic illustration of the energy levels of a quantum dot connected to a source and drain electrode through a tunnel junction. The voltage between the source and dot is given by $V$. In the left figure the number of electrons in the dot is fixed at $N$ and the tunnel current vanishes (Coulomb blockade). In the right figure, a single electron tunneling occurs from the source to drain electrode because the charging energy for the dot is nearly the same for $N$ and $N+1$ electrons.

to a term proportional to $N$ in the electrostatic energy. Therefore, the total electrostatic energy is given by

$$E_N = \frac{Q^2}{2C} - \alpha e Q V_G, \tag{4.1}$$

where $Q = -eN$, the coefficient of the term proportional to $Q^2$ has been expressed in terms of capacitance $C$, and that of the term proportional to $Q$ by gate voltage $V_G$ with $\alpha$ being an appropriate numerical constant.

### 4.2 Coulomb Blockade and Single Electron Tunneling

Consider the tunnel current through the antidot sandwiched by a source and drain electrode through a thin tunnel barrier. Figure 21 gives an illustration of the energy diagrams. In the left figure a nonvanishing energy is required in moving one electron from the source electrode to the dot or from the dot to the drain electrode, i.e., the number of electrons in the dot is fixed at $N$. Therefore, the tunneling process is prohibited at sufficiently low temperatures. This is called the Coulomb blockade.[38-41]

When the gate voltage is appropriate, we have $\Delta U_N = 0$ and the energy of the dot containing $N$ and $N+1$ electrons becomes the same. In this case an electron tunnels into the dot from the source electrode and then tunnels out into the drain electrode. The electron tunneling out to the drain needs not be the same as the electron tunneling into the dot from the source. While the dot contains an extra electron or charge and until the extra charge disappears from the dot, no other electrons can enter the dot because of the charging energy. Therefore, this tunneling process is called a single electron tunneling.

Figure 22 gives the behavior of the conductance in the plane consisting of the source-drain voltage $V$ and the gate voltage $V_G$. In the limit $V \to 0$, a series

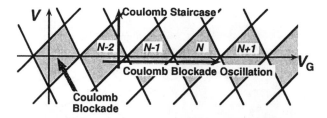

**Fig. 22** A kind of the phase diagram in the plane of the source-drain voltage $V$ and the gate voltage $V_G$. In the dirk region the tunneling current vanishes due to the Coulomb blockade. When $V_G$ is varied for a fixed $V$ a series of sharp peaks equally spaced called a Coulomb blockade oscillation. When $V$ is varied for a fixed $V_G$, a Coulomb staircase is observed.

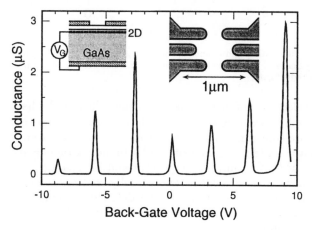

**Fig. 23** An example of experimental results of a Coulomb blockade oscillation observed in a quantum dot fabricated by a split-gate technique. The inset shows the structure of a single heterojunction with a back gate (left) and the structure of the split gate (right).

of sharp peaks appear at the discrete gate voltages due to the single electron tunneling. When $V$ is increased slightly, the region where the tunneling is possible widens in proportion to $V$. Consequently the plane is split into various regions given by two kinds of parallel lines. In the dirk region the tunneling is prohibited due to the Coulomb blockade and in the white region the tunneling is possible. We have a Coulomb blockade oscillation as a function of $V_G$ for a given $V$ and a Coulomb staircase as a function of $V$ for a given $V_G$.

Figure 23 shows an example of a Coulomb blockade oscillation observed in quantum dots fabricated at a GaAs/AlGaAs heterostructure by a split-gate technique.[42] Electrons are confined into a dot by a negative voltage on the gate and the tunneling current flows below the narrow spacing from the bottom to

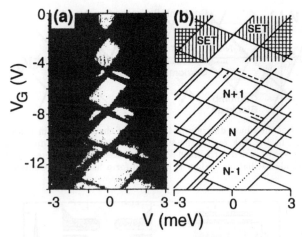

**Fig. 24** The differential conductance in the $V$-$V_G$ plane (left). The conductance is larger in the dirk region and smaller in the white region. The various structures observed are shown in the right figure.

the top. The electron number in the dot is controlled by the back gate.

The Coulomb blockade and the single electron tunneling are nothing so new. In fact they are sometimes observed in conventional tunnel junctions when a barrier insulator contains an impurity and a tunneling takes place through the localized level.[43,44] It is quite an important and interesting development that we can now fabricate such systems and actually control the motion of a single charge. There are various physics in the detailed behavior of the Coulomb blockade oscillation such as the change in the height of the peak and a slight change in the period.

Figure 24 gives observed conductance in a gray scale in the plane consisting of the gate and source-drain voltage. The white Coulomb-blockade region is apparent. In the dirk region where a single electron tunneling occurs, various fine structures are observed. For example, the conductance sometimes becomes smaller for a larger source-drain voltage. This is due to the quantization of energy levels in quantum dots. For a large source-drain voltage an electron can tunnel into an excited energy levels which may have a large probability for tunneling into the dot from the source but a small probability for tunneling out to the drain electrode. In the regime of the single-electron tunneling, this process tends to block the current because of the Coulomb blockade.

## 4.3 Single Electron Turnstile

Numerous experiments have been performed in connection to this single electron tunneling and various new devices have been proposed. Figure 25(a) gives

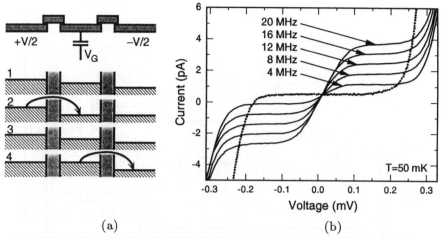

**Fig. 25** A schematic illustration of a turnstile device, the principle of the operation, and an example of observed experimental results. The current is quantized into a value proportional to the frequency in the wide region of the source-drain voltage.

the structure of a turnstile device.[45] We first apply a certain source-drain voltage in such a way that the tunnel current is prohibited through both quantum dots due to the Coulomb blockade. When we lower the voltage of the central region, an electron tunnels into it from the left electrode but cannot tunnel out to the right electrode because of the Coulomb blockade of the right quantum dot. When we raise the voltage, the extra charge $-e$ stored in the central region tunnels out into the right electrode. In this way the charge $-e$ moves from left to right per cycle. The current in this turnstile is determined by the elementary charge $e$ and the frequency of the voltage modulation of the central region.

Figure 25(b) shown an example of observed experimental results. The current is proportional to the modulation frequency for 4, 8, 12, ... MHz. The turnstile is considered to be a possible current-standard.

## 5   History and Outlook

This research field has a history of only thirty years and is closely related to the development of the LSI technology of the semiconductor industry and the crystal growth technique such as MBE and MOCVD as is illustrated in Fig. 26. In fact, the present 2D physics was born together with MOSFET's (MOS field effect transistors), where the integer quantum Hall effect was discovered for the first time. In a sense, it is an interdisciplinary research field where strong interactions among basic science, device application, and technology

36

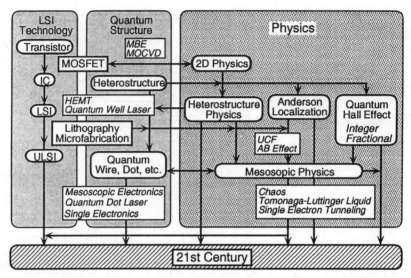

**Fig. 26** A schematic illustration of the relation between different fields related to present mesoscopic physics.

play dominant roles.

The 2D physics expanded quite a bit when the quality of semiconductor heterostructures was improved and the fractional quantum Hall effect was discovered. The heterostructure itself has provided also various fundamental problems including the band alignment and the connection of the electron wave at interfaces. The heterostructure is now being used as real devices such as quantum-well lasers and HEMT, which is the high electron mobility transistor used in satellite broadcasting.

The reduction in the size of semiconductor devices has lead to the development in the lithography and microfabrication. The combination of this lithography and crystal growth has lead to quantum wires, dots, antidots, and so on. These new quantum structures have contributed greatly to the further expansion of the research field. Observing such a history, it is very difficult to predict its future direction exactly. It is certainly correct to say at least that this field will continue to grow further and further in next twenty years.

### Acknowledgments

This work is supported in part by Grant-in-Aid for Scientific Research on Priority Area "Mesoscopic Electronics: Physics and Technology" from the Ministry of Education, Science and Culture.

# References

1. T. Ando, A.B. Fowler, and F. Stern, *Rev. Mod. Phys.* **54**, 437 (1982).
2. K. von Klitzing, G. Dorda, and M. Pepper, *Phys. Rev. Lett.* **45**, 494 (1980).
3. D.C. Tsui, H.L. Stormer, and A.C. Gossard, *Phys. Rev. Lett.* **48**, 1559 (1982).
4. B.J. van Wees, H. van Houten, C.W.J. Beenakker, J.G. Williamson, L.P. Kouwenhoven, D. van der Marel, and C.T. Foxon, *Phys. Rev. Lett.* **60**, 848 (1988).
5. D.A. Wharam, T.J. Thornton, R. Newbury, M. Pepper, H. Ahmed, J.E.F. Frost, D.G. Hasko, D.C. Peacock, D.A. Ritchie, and G.A.C. Jones, *J. Phys. C* **21**, L209 (1988).
6. Y. Takagaki, K. Gamo, S. Namba, S. Ishida, S. Takaoka, K. Murase, K. Ishibashi, and Y. Aoyagi, *Solid State Commun.* **68**, 1051 (1989).
7. M.L. Roukes, A. Scherer, S.J. Allen, Jr., H.G. Craighead, R.M. Ruthen, E.D. Beebe, and J.P. Harbison, *Phys. Rev. Lett.* **59**, 3011 (1987).
8. H.U. Baranger and A.D. Stone, *Phys. Rev. Lett.* **63**, 414 (1989).
9. C.W.J. Beenakker and H. van Houten, *Phys. Rev. Lett.* **63**, 1857 (1989).
10. H.U. Baranger, D.P. DiVincenzo, R.A. Jalabert, and A.D. Stone, *Phys. Rev. B* **44**, 10637 (1991).
11. H. Akera and T. Ando, *Phys. Rev. B* **41**, 11967 (1990).
12. R. Landauer, *IBM J. Res. Dev.* **1**, 223 (1957); *Philos. Mag.* **21**, 863 (1970).
13. R. Kubo, *J. Phys. Soc. Jpn.* **12**, 570 (1957).
14. P.A. Lee and D.S. Fisher, *Phys. Rev. Lett.* **47**, 882 (1981).
15. D.J. Thouless, *Phys. Rev. Lett.* **47**, 972 (1981).
16. D.C. Langreth and E. Abrahams, *Phys. Rev. B* **24**, 2978 (1981).
17. M. Büttiker, *Phys. Rev. Lett.* **57**, 1761 (1986); *IBM J. Res. Dev.* **32**, 317 (1988).
18. M. Okada, M. Saito, K. Kosemura, T. Nagata, H. Ishiwari, and N. Yokoyama, *Superlatt. Microstr.* **10**, 493 (1991).
19. M. Okada, M. Saito, M. Takatsu, .E. Schmidt, K. Kosemura, and N. Yokoyama, *Semicond. Sci. Technol.* **7**, B223 (1992).
20. T. Ando, *Phys. Rev. B* **44**, 8017 (1991).
21. D.R. Hofstadter: *Phys. Rev. B* **14**, 2239 (1976).
22. D. Weiss, M.L. Roukes, A. Menschig, P. Grambow, K. von Klitzing, and G. Weimann, *Phys. Rev. Lett.* **66**, 2790 (1991).
23. R. Fleischmann, T. Geisel, and R. Ketzmerick, *Phys. Rev. Lett.* **68**, 1367 (1992).
24. E.M. Baskin, G.M. Gusev, Z.D. Kvon, A.G. Pogosov, and M.V. Entin, *JETP Lett.* **55**, 678 (1992).
25. T. Nagao, *J. Phys. Soc. Jpn.* **64**, 4097 (1995).
26. T. Nagao, *J. Phys. Soc. Jpn.* **65**, 2606 (1996).
27. F. Nihey and K. Nakamura, *Physica B* **184**, 398 (1993).
28. D. Weiss, K. Richter, A. Menschig, R. Bergmann, H. Schweizer, K. von Klitzing, and G. Weimann, *Phys. Rev. Lett.* **70**, 4118 (1993).

38

29. S. Ishizaka and T. Ando, Phys. Rev. B (submitted for publication).
30. S. Ishizaka, F. Nihey, K. Nakamura, J. Sone, and T. Ando, Jpn. J. Appl. Phys. **34**, 4317 (1995).
31. S. Ishizaka, F. Nihey, K. Nakamura, J. Sone, and T. Ando, Phys. Rev. B **51**, 9881 (1995).
32. K. Nakamura, S. Ishizaka, and F. Nihey, Physica B **197**, 144 (1994).
33. H. Silberbauer and U. Rössler, Phys. Rev. B **50**, 11911 (1994).
34. S. Uryu and T. Ando, Jpn. J. Appl. Phys. **34**, 4295 (1995).
35. S. Uryu and T. Ando, Phys. Rev. B **53**, 13613 (1996).
36. S. Uryu and T. Ando, Physica B **227**, 138 (1996).
37. S. Uryu and T. Ando, 23rd International Conference on Physics of Semiconductors, edited by M. Scheffler and R. Zimmermann (World Scietific, Singapore, 1996), p. 59.
38. D.V. Averin and K.K. Likharev, J. Low Temp. Phys. **62**, 345 (1986).
39. D.V. Averin and K.K. Likharev, Mesoscopic Phenomena in Solids, edited by B.L. Altshuler, P.A. Lee, and R.A. Webb (Elsevier, Amsterdam, 1991), p. 173.
40. M.A. Kastner, Rev. Mod. Phys. **64**, 849 (1992).
41. M.A. Kastner, Phys. Today **46**, 24 (1993).
42. J. Weis, R.J. Haug, K. von Klitzing, and K. Ploot, Phys. Rev. Lett. **71**, 4019 (1993).
43. R.I. Shekhter, Sov. Phys. JETP **36**, 747 (1973).
44. I.O. Kulik and R.I. Shekhter, Sov. Phys. JETP **41**, 308 (1975).
45. L.J. Geerligs, V.F. Anderegg, P.A.M. Holweg, J.E. Mooij, H. Pothier, D. Esteve, C. Uribina, and M.H. Devoret, Phys. Rev. Lett. **64**, 2691 (1990).

# Nonequilibrium Physics and the Origins of Complexity in Nature[a]

J. S. Langer

*Physics Department, University of California*
*Santa Barbara, CA 93106*

Complex spatial or temporal patterns emerge when simple systems are driven from equilibrium in ways that cause them to undergo instabilities. Dendritic (i.e. snowflake-like) crystal growth is an especially clear example. We now have a well tested and remarkably detailed theory of how dendrites form and grow. We understand the role of intrinsically small effects such as crystalline anisotropy, and we even can trace the growth of dendritic sidebranches all the way back to their origins as microscopic thermal fluctuations. This extreme sensitivity to perturbations explains many long-standing puzzles about snowflakes. A similar sensitivity is emerging in the physics of dynamic fracture, a phenomenon which is similar in some ways to dendritic growth but which now seems to be governed by entirely different mechanical and mathematical principles. Extreme sensitivity to perturbations and system parameters is likely to be a characteristic feature of pattern-forming systems in general, including those that occur in biology. If so, we shall not succeed in reducing the physics of complexity to a small number of universality classes. On the contrary, we shall have to be prepared for a wide variety of challenges and surprises in this field.

I cannot remember a time when I have felt more optimistic about the intellectual vitality of physics than I do now. I am cautiously optimistic about the remarkable developments in particle theory and cosmology, but those are too far from my areas of expertise for me to say much about them in public. What impresses me more is that we are on the verge of deep new understanding of a wide range of every-day phenomena, most of which appear to be "complex" in some sense of that word. For the first time in history, we have the tools — the experimental apparatus and the computational and conceptual capabilities — to answer questions of a kind that have always caught the imaginations of thoughtful human beings.

I want especially to emphasize my conviction that the physics of complex, every-day phenomena will be an extraordinarily rich and multi-faceted part of the science of the next Century. It may, however, have to be pursued in a manner that is different from that of physics in the 20th Century. It will be strongly interdisciplinary, of course. We physicists shall have to work in close collaboration with many kinds of scientists and engineers, often in areas where

---

[a]Lecture presented at the Princeton 250th Anniversary Conference on Critical Problems in Physics, October 31, 1996, and at the 11th Nishinomiya-Yukawa Memorial Symposium on Physics in the Twenty-First Century, November 7, 1996.

we have not been the pioneers. That will be a major challenge for us and our institutions.

It is even more important, I think, to recognize that we shall have to modify our innate urge to speculate about unifying principles at very early stages of our research projects, often long before we have dug deeply into the details of specific phenomena. Here I differ from some of my colleagues. For reasons that I shall try to explain, I suspect that complex systems will not fall usefully into a small set of universality classes. On the contrary, I think that our glimpses so far into the physics of complex systems indicate that this world is larger than we had expected, and that we may be much closer to the beginning than to the end of this chapter in the history of science.

About twenty-five years ago, P.W. Anderson wrote an essay entitled "More Is Different."[1] He meant that systems with large numbers of degrees of freedom often behave in novel and surprising ways. In his words: "...at each [increasing] level of complexity new properties appear, and the understanding of the new behaviors requires research which ... is as fundamental in its nature as any other." He was right, as usual, and quite early in arguing this point. I think, however, that even Anderson may have underestimated how many fundamentally new directions for investigation would emerge, and how profoundly that diversity of challenges may change the nature of research in physics.

Biology, of course, comes first on my list of new areas of inquiry for physicists in the next Century. Life on Earth may in some sense be the most complex phenomenon in the universe — a sobering thought with profound philosophical as well as scientific implications. I think that we must place the understanding of life, and human life in particular, at the top of the list of our most compelling intellectual challenges. Unfortunately, I am no more competent to talk about biology than I am to say anything sensible about elementary particles. However, the subject that I shall address, nonequilibrium physics and the origins of complexity in nature, moves us in the direction of thinking about the physics of biological systems.

My favorite way to illustrate the impact of new developments in nonequilibrium physics is to talk about snowflakes and dendritic crystal growth. This is a classic problem in the theory of pattern formation, one with a long history that is relevant to th e issues I want to raise. I also shall make some brief remarks about fracture of solids and earthquake dynamics, primarily to emphasize the scope and diversity of research in this field even within the conventional bounds of non-biological materials phy sics.

*Kepler's Snowflakes*

In 1611, Johannes Kepler published a monograph called "The Six-Cornered Snowflake."[2] Kepler wrote it as a "New Year's Gift" for one of his patrons in

Prague. It was republished in 1966 with commentary by the historian and philosopher Lancelot Law Whyte, who pointed out that it marked the first recorded instance in which morphogenesis had been treated as a subject for scientific rather than theological discussion.

In his essay, Kepler wondered how such beautifully complex and symmetric structures can emerge, literally, out of thin air. It would be easy to understand, he thought, if these were animate objects that had souls and God-given purposes for being, and wh ich therefore possessed an innate force for growth and development. But no such option seemed available. He went on to speculate about natural geometric ways to generate six-fold symmetries and, long before he had any reason to believe in the existence of atoms, he drew pictures of hexagonal close-packed arrays of spheres. But he concluded that this was a problem that would have to wait for future generations of scientists; he did not have the tools to solve it in his own time. (I wonder whether my patrons, the DOE or the NSF, would be satisfied with a progress report in which I admitted that I couldn't solve the problems that I had posed!)

*Thoreau's Snowflakes*

About two and a half centuries after Kepler (on January 5, 1856), the American essayist and naturalist Henry David Thoreau[3] described in his Journal the snowflakes that he observed while walking in the woods, and then remarked: "How full of the creative genius is the air in which these are generated! I should hardly admire them more if real stars fell and lodged on my coat. Nature is full of genius, full of the divinity; so that not a snowflake escapes its fashioning hand." Like Kepler, Thore au spoke of the "genius" of nature, and he could not resist using the term "divinity."

Kepler's and Thoreau's questions, and ours to this day, are very deep. We still ask how snowflakes form. What causes their six-fold symmetry? Why the treelike structure? Why do their decorative features — sidebranches, tip splittings, bulges, and the l ike — occur at the same places on each of the six branches? Do the growing parts of these branches somehow communicate with each other? And why does it seem that no two snowflakes are ever exactly alike?

*Metallurgical Microstructures*

By the 1960's, when Whyte was writing his comments about Kepler, the crystallographers and the atomic and solid-state physicists had figured out, at least in principle, why ice crystals have intrinsic hexagonal symmetry. At about the same time, however, the metallurgists were realizing that crystallography is only a small part of the story. At best, crystallography can predict the *equilibrium* form of the material which, in the case of ice, is a compact crystal with hexagonal symmetry, but with no branches or sidebranches or any of the complex structure of snowflakes. Crystallography, plus equilibrium thermodynamics and statistical mechanics, provides no clue about how to predict and control the microstructures of cast alloys, or welds, or any of the vast number of examples in nature where complex patterns emerge spontaneously in relatively simple materials.

A solidified metallic alloy, when etched and examined through a microscope, often looks like a forest of overdeveloped snowflakes with generations upon generations of sidebranches. During solidification, the molten material crystallizes in much the same way that water vapor crystallizes in the atmosphere, and the resulting "dendritic" or tree-like structure — the so-called "microstructure" of the alloy — is what determines its mechanical and electrical properties. To predict microstructures, metallurgists need to know the growth rates and characteristic length scales that occur during the transient processes of solidification. They need especially to know how those speeds and lengths are determined by constitutive properties of the materials such as thermal conductivity, surface tension, etc., and also by growth conditions such as the temperature and composition of the melt.

*Solidification as a Free-Boundary Problem*

The dendrite problem is easiest to describe if we restrict our attention to the case of solidification of a pure fluid held at a temperature slightly below its freezing point.[4] We must also, at this point, turn our attention to systems that are intrinsically simpler than real snowflakes made out of water molecules. Most of the interesting metallurgical materials are much easier than water for theoretical purposes.

Imagine a piece of solid completely immersed in its undercooled melt. Thermodynamics tells us that the free energy of the system decreases as the solid grows into the liquid. For this to happen, the latent heat generated during growth must be carried away from the solidification front by some transport mechanism such as diffusion. The motion of the front is governed by the interplay between two simple and familiar processes — the irreversible diffusion of heat and the reversible work done in the forma tion of new surface area.

In mathematical language, we say that the shape of the emerging crystal is the solution of a free-boundary problem. That problem consists of a diffusion equation for the temperature field $u$,

$$\frac{\partial u}{\partial t} - D\nabla^2 u = 0;$$ (1)

the condition of heat conservation at the moving solidification front:

$$v_n = -D \left[\frac{\partial u}{\partial n}\right]_{\text{liquid}} + D \left[\frac{\partial u}{\partial n}\right]_{\text{solid}} ; \qquad (2)$$

and the Gibbs-Thomson condition for the temperature $u_s$ at a curved interface:

$$u_{\text{interface}} = -d_0 \kappa. \qquad (3)$$

Here, the dimensionless field $u$ is the difference between the local temperature and the bulk freezing temperature, measured in units of the ratio of the latent heat to the specific heat. At the boundaries of the system, very far from the solidification front, $u = u_\infty < 0$; where $|u_\infty|$ is the dimensionless undercooling that plays the role of the driving force for this nonequilibrium system. $D$ is the thermal diffusion constant; $n$ denotes distance along the normal directed outward from the solid; $v_n$ is the normal growth velocity; $d_0$ is a "capillary length" — ordinarily of order Angstroms — that is proportional to the liquid-solid surface tension; and $\kappa$ is the sum of the principal curvatures of the interface. It is important that the surface tension may depend on the orientation of the interface relative to the axes of symmetry of the crystal.

*Shape Instabilities and Singular Perturbations*

The key to understanding how such a simply posed problem can have complex dendritic solutions is the observation that smooth shapes generated by these equations are intrinsically unstable. Such instabilities appear throughout physics — in the lightning-rod effect, for example, or in the stress concentrations at the tips of the cracks that I shall mention later. In the case of dendritic crystal growth, the slightest bump on the surface of an otherwise smooth solidification front concentrates the diffusion gradients and increases the heat flow from the surface, thus enhancing the local growth rate. The small bump grows out into a dendrite; the primary dendrite undergoes secondary instabilities that lead to sidebranches; and thus a complex pattern emerges whose shape is highly sensitive to the precise way in which these instabilities have occurred.

There is a natural length scale associated with this instability. A solidification front growing at speed $v$ is unstable against deformations whose wavelengths are larger than

$$\lambda_s = 2\pi\sqrt{\frac{2Dd_0}{v}}. \qquad (4)$$

Note that $\lambda_s/2\pi$ is the geometric mean of a macroscopic length, the diffusion length $2D/v$, and the microscopic length $d_0$. It provides a first, order-of-magnitude estimate for the length scales observed in dendritic growth. In particular, $\rho$, the radius of curvature of the dendritic tip, scales like $\lambda_s$. Thus the ratio of $\lambda_s$ to $\rho$, conventionally defined by the parameter

$$\sigma^* \equiv \left(\frac{\lambda_s}{2\pi\rho}\right)^2 = \frac{2Dd_0}{v\rho^2}, \tag{5}$$

has been found experimentally to be very nearly constant for a given material over a wide range of growth conditions.

The problem of obtaining a first-principles estimate of $\sigma^*$ remained unsolved, however, until about ten years ago. The crucial observation was that the capillary length $d_0$, and therefore $\sigma^*$ itself, is a singular perturbation whose presence, no matter how small, qualitatively changes the mathematical nature of the equations. For $d_0 = 0$, there is a continuous family of steady-state solutions. For any nonzero $d_0$, however, there is only at most a discrete set of solutions, one of which may (or may not) be a stable attractor for this dynamical system. Moreover, the attracting solution looks like a regular dendrite only if there is some crystalline anisotropy in the surface tension. The parameter $\sigma^*$ determines growth rates and tip radii in much the same way we had guessed from stability theory, but it is a function of the anisotropy, and it vanishes when the anisotropy strength goes to zero. In the latter case, when the surface tension is isotropic, the patterns are irregular and even more complex. They may, in some limiting cases, become truly fractal. In short, complex dendritic patterns are governed by a weak correction — the anisotropy — to a weak but singular perturbation — the surface tension.

*Thermal Fluctuations and Sidebranches*

There are two, more recent, developments in the dendrite story whose implications are relevant to the issues I want to raise here. The first has to do with the origin of sidebranches. Consider a small perturbation at the dendrite tip, caused perhaps by an impurity in the melt or a thermal fluctuation. The latest theories predict that such a perturbation grows like

$$\zeta(s) \approx \zeta(0) \exp\left[+\frac{\text{const.}}{\sqrt{\sigma^*}}\left(\frac{s}{\rho}\right)^b\right] \tag{6}$$

where $\zeta(s)$ is the size of the deviation from a smooth steady-state shape at a distance $s$ from the tip of the dendrite. The exponent $b$ is $1/4$ in an approximation in which the dendrite is cylindrically symmetric[5] When realistic axial anisotropy is taken into account, the relevant exponent is $b = 2/5$[6] (This equation is typical of the results that emerge from the singular perturbation theory. Note the special role of small values of $\sigma^*$ in this formula. It is impossible to approximate $\zeta$ by a series expansion in powers of $\sigma^*$.) The bump that is formed from this deformation grows out at a fixed position in the laboratory frame of reference; it becomes a sidebranch.

A natural question to ask is: How strong are the initial perturbations $\zeta(0)$ that produce the sidebranches seen in the real world? Might thermal fluctuations be strong enough? We can answer this question by using formulas like (6) to predict the way in which the first sidebranches appear behind the tip. We now know, thanks to recent work along these lines by Brener, Bilgram, and their colleagues,[6,7,8,9] that sidebranches in the most carefully studied three-dimensional dendritic systems are indeed formed by selective amplification of thermal fluctuations. I find it quite remarkable that this pattern-forming process is so sensitive that we can trace macroscopic structures all the way back to their origins as fluctuations on the scale of molecular motions.

*A Field-Theoretic Description of Solidification Patterns*

The last part of the dendrite story that I want to mention is the so-called "phase-field model." The free-boundary problem described in (1)-(3) looks esoteric to most theoretical physicists who, like myself, grew up immersed in field theories. It happens, however, that it is easy and perhaps even useful to rewrite these equations in a field-theoretic language. To do this, add a source to the thermal diffusion equation (1):

$$\frac{\partial u}{\partial t} - D\nabla^2 u = -\alpha \frac{\partial \phi}{\partial t}. \tag{7}$$

Here, $\alpha$ is a constant that is proportional to the latent heat, and $\phi$ is a "phase field" — i.e. a local order parameter — that tells us, by being near one or the other of two equilibrium values that we shall fix for it, whether the system is in its liquid or solid phase. According to (7), heat is generated or absorbed at a moving liquid-solid interface when, at such a point, $\phi$ is changing from one of its equilibrium values to the other.

Eq. (7) must be supplemented by an equation of motion for $\phi$, which I shall write in a specially simple but suggestive form:

$$-\frac{1}{\Gamma}\frac{\partial \phi}{\partial t} = -\nabla^2 \phi - \mu^2 \phi + \lambda \phi^3 + \beta u. \tag{8}$$

This is the familiar time-dependent Ginzburg-Landau equation with an explicit minus sign in front of $\mu^2\phi$, indicating that we are in the symmetry-broken low-temperature state of the system. The two stable values of $\phi$ are approximately $\pm\mu/\sqrt{\lambda}$. As long as $\mu$ is large, so that the interface between the two phases is thin, Eqs. (7) and (8) reduce to Eqs. (1)-(3), in this case with a nonzero but purely isotropic surface tension. The parameters $\alpha$, $\beta$, $\Gamma$, $\lambda$, and $\mu$ can be evaluated in terms of the parameters that appear in the earlier equations.

The phase-field model has been studied extensively in recent years.[10] It even has been used to confirm the singular-perturbation theory of dendritic pattern selection and to compute the parameter $\sigma^*$ defined in (5).[11] Although (8) is a very stiff partial differential equation for large $\mu$, the combination of (7) and (8) lends itself to numerical analysis more easily than the equivalent free-boundary problem. My main reason for mentioning this model, however, is to point out that a familiar-looking field theory of this kind may produce patterns of unbounded complexity. These patterns are exquisitely sensitive to small perturbations, for example, the addition of crystalline anisotropy or external noise. They are also "chaotic," in the technical meaning of the word, in their sensitivity to initial conditions. The fact that such complex behavior can occur here must be kept in mind when studying similar equations in circumstances where reliable experimental or numerical tests are not available.

### The Symmetry and Diversity of Snowflakes

Recent theoretical and experimental developments finally have provided answers to our age-old questions about snowflakes. We now understand that they nucleate as tiny crystals in a supersaturated mixture of water vapor and air, and that the intrinsic hexagonal symmetry of ice guides the subsequent morphological instability and produces six nearly identical arms. The temperature and humidity of the air that surrounds each growing snowflake are very nearly constant on the relevant length scales of millimeters or less; thus all the arms respond in much the same way to their surroundings, and the precision and sensitivity of the controlling mechanisms make it unnecessary for there to be any direct communication between the distant parts of the emerging structure in order for them to grow in essentially identical ways.

All of this is happening in a turbulent atmosphere. The characteristic length scales of atmospheric turbulence, however, are meters, or kilometers — enormously larger than the sizes of the snowflakes themselves. As the snowflakes are advected through this environment, they encounter changing growth conditions. Sometimes their arms freeze quickly in regions that are particularly cold; at other times they grow slowly or even melt back as they encounter warmer regions. Our 20th Century understanding of turbulence tells us that the trajectories of neighboring snowflakes must diverge, that each flake must encounter its own unique sequence of growth conditions. In this technical sense, no two snowflakes are alike.

### Fracture Dynamics

The dendrite problem seemed "mature" thirty years ago, but we know now that it was not. It did mature greatly in the last decade and, although there remain many important unsolved problems, it may have peaked as a center of intellectual excitement. To counter any notion that the field of nonequilibrium pattern formation as a whole is reaching maturity, I want to describe briefly one other supposedly mature area of research that is poised for major fundamental developments. This is the broad area of mechanical failure of solids, fracture in particular.

For obvious reasons, human beings must have become interested in fracture long before they started to ask philosophical questions about snowflakes. Similarly, the modern field of fracture mechanics — the study of when things break — was a well established specialty within engineering and applied mathematics before the dendrite problem had even been stated in precise terms. Remarkably, however, fracture *dynamics* — the study of how cracks move — is literally *terra incognita*, largely, I suppose, because the engineer's responsibility is to prevent such catastrophes before they start, but also because propagating fracture has been very hard to study either experimentally or theoretically until now.

I first started looking at dynamic fracture because I thought that a crack moving through a brittle solid should have much in common theoretically with a dendrite. The two phenomena look quite similar, at least superficially. Both are nonequilibrium, finger-like structures that move under the influence of external forces. The elastic stress concentration at the tip of a crack occurs for much the same reason that the diffusion flux is concentrated near the tip of a dendrite. Both exhibit some kind of sidebranching behavior. The scientific problem, in both cases, is to predict the propagation speed, the characteristic length scales, and the stability of the patterns that are formed. I also became interested in the dynamics of fracture on earthquake faults, and I wondered whether this behavior is an example of "self-organized criticality."[12] It is not, but it does come close. The underlying physics of fracture, on the other hand, turns out to be entirely different from the physics of dendrites.[13] My education in dendrite theory has been of very little use for studying fracture, but I have learned a great deal that is new to me.

One of the first things I learned about fracture is how many obvious and ostensibly simple questions remain unanswered. For example, we do not know in any really fundamental way how to predict whether failure will occur in a solid via brittle fracture or ductile yielding. There are elegant and, at least to some extent, successful theories involving the creation and motion of dislocations. But all the same phenomena occur in amorphous materials, where dislocations have no meaning. So there must be more to learn here. In fact, we still have no first-principles theory of the glassy state, and we certainly have no fundamental understanding of failure mechanisms in glasses. The state of affairs regarding granular materials is even more wonderfully uncertain.

My recent research has focussed on the stability of dynamic fracture and on the question of how — sometimes — complex patterns are generated during the failure of solids. Here again, the open questions have to do with familiar phenomena. We generally think of fracture surfaces as being intrinsically rough, perhaps even fractal. But we also know that they can be very smooth as, for example, in the case of a slow crack in an automobile windshield, or a sharp cleavage surface on a prehistoric stone tool. What is the nature of the instability that determines whether a crack will produce a smooth or a rough surface? Is this the same instability that causes an earthquake fault zone to be an intricate, interconnected network of fractures that extends for tens of kilometers both laterally and vertically in the neighborhood of a main fault?[14] Is this geometric complexity the reason for the broad, Gutenberg-Richter distribution of earthquake magnitudes, or does the chaotic dynamics of individual fault segments also play a role?[15]

I think I now know at least the broad outlines of answers to some of these questions. Recent experiments by Fineberg *et al*[16,17] have made it clear that bending instabilities at high speeds are intrinsic features of fracture; they are not just artifacts of special experimental conditions. My recent work with Ching and Nakanishi[18] leads me to be fairly certain that this instability is a general property of stress fields near the tips of moving cracks. The "lightning-rod" effect does play a crucial role here in concentrating stresses near the crack tip, but the crucial ingredient is the "relativistic" response of an elastic solid when the crack tip is moving at an appreciable fraction of the sound speed. There is no analog of such an effect in dendritic crystal growth. So far as I can see, the fracture problem and the dendrite problem are in two entirely different universality classes.

If I am correct in these assertions, then fracture dynamics is at roughly the same state as was solidification theory thirty years ago. We know (maybe) what triggers complex behavior in fracture, but little or nothing about mechanisms that may control the underlying instability. Our new results imply that, just as in the dendrite problem, the dynamics is strongly sensitive to the specifics of the models, in this case to the detailed mechanisms of cohesion and decohesion at crack tips or stick-slip friction on earthquake faults. There are even hints that some of these mechanisms may be singular perturbations whose presence makes qualitative changes in the system. But this is a subject for another occasion.

### The Physics of Complexity

I hope that it is clear by now where I have been going in this discussion. Complex patterns are produced during crystal growth or fracture because systems are driven from thermal or mechanical equilibrium in ways that cause them to become unstable. Departure from equilibrium, external driving forces, and instability are important, fundamental elements of our understanding of these classes of phenomena. But those general elements tell us only a small part of what we need to know. A deep scientific understanding requires that we find out what mechanisms control the instabilities and that we learn in detail how the interplay between those mechanisms and the driving forces produces the patterns that we see. Because those mechanisms are acting in intrinsically unstable situations, dynamical systems may be strongly sensitive to them. Their effects may be highly specific. The patterns in some extreme cases may be completely regular, or fractal, or fully chaotic; but those are just special limits. Between those limits, the range of possibilities is enormous, often bewilderingly so. Whether we like it or not, that is where we must look to understand the real world.

What, then, are the prospects for a new "science of complexity"? Quite good, I think, if our measure of goodness is the variety and scope of the intellectual challenges that we shall encounter. Note that my scale of goodness is different from the traditions of 20th Century physics, which insist on grand unifications and underlying simplicity. I am perfectly comfortable with the idea that we may never find a set of unifying principles to guide us, in the manner of Gibbsian equilibrium statistical mechanics, to solutions of problems in the nonequilibrium physics of complex systems. We seem to be discovering far too large a variety of intrinsically different behaviors for such a synthesis to be possible. Nor, as I have said, do I think that we shall discover that all complex systems can be sorted into some small set of universality classes. That is why I find it hard to be enthusiastic about modelistic searches for generic behavior, usually computational experiments, that are not tightly coupled to real physical situations.

This is not to say, however, that I am pessimistic about the discovery of new basic principles. Just the opposite — I am sure that they will appear, and that they will do so in ways that will surprise us. We already have one example of a new idea in "self-organized criticality."[12] Whether or not "SOC" matures into a precise predictive mode of analysis, I believe that it will persist as a provocative way of thinking about many complex systems. Other such insights, perhaps pertaining to more specific classes of phenomena, or to phenomena of a kind that we have never seen before, may be just around the corner. In a field so rich with new experimental and observational opportunities, and with the huge world of biological systems just beginning to be explored from this point of view, it seems obvious that new unifying concepts will emerge as well as a vast amount of new physics.

50

## Acknowledgments

The research on which this paper is based was supported by DOE Grant No. DE-FG03-84ER45108 and NSF Grant No. PHY-9407194. I especially want to thank J.M. Carlson and S.A. Langer for constructive critiques of early versions of this manuscript, and Elizabe th Witherell for pointing me toward Thoreau.

## References

1. P.W. Anderson, *Science* **177**, 393 (1972).
2. J. Kepler, *The Six-Cornered Snowflake*, translated by C. Hardie, with accompanying essays by B. J. Mason and L. L. Whyte (Clarendon Press, Oxford, 1966) [originally published as *De Nive Sexangula* (Godfrey Tampach, Frankfurt am Main, 1611)].
3. Henry D. Thoreau, *The Journal of Henry David Thoreau*, 14 volumes, edited by Bradford Torrey and Francis Allen (Houghton Mifflin, Boston, 1906; Peregrine Smith, Salt Lake City, 1984) Volume 8, pp. 87-88.
4. I can make no attempt here to cite detailed references to the literature in solidification physics. A few reviews that may be useful are the following: J.S. Langer, *Rev. Mod. Phys.* **52**, 1 (1980); J.S. Langer, in *Chance and Matter*, proceedings of the Les Houches Summer School, Session XLVI, edited by J. Souletie, J. Vannimenus, and R. Stora (North Holland, Amsterdam, 1987), p. 629; D. Kessler, J. Koplik and H. Levine, *Adv. Phys.* **37**, 255 (1988); P. Pelcé, *Dynamics of Curved Fronts* (Academic Press, New York, 1988).
5. J.S. Langer, *Phys. Rev.* A **36**, 3350 (1987).
6. E. Brener and D. Temkin, *Phys. Rev.* E **51**, 351 (1995).
7. M. Ben Amar and E. Brener, *Phys. Rev. Lett.* **71**, 589 (1993).
8. E. Brener, *Phys. Rev. Lett.* **71**, 3653 (1993).
9. U. Bisang and J.H. Bilgram, *Phys. Rev. Lett.* **75**, 3898 (1995).
10. J.A. Warren and W.J. Boettinger, *Acta Metall. et Mater.* **43**, 689 (1995).
11. A. Karma and W.-J. Rappel, *Phys. Rev. Lett.* **77**, 4050 (1996).
12. P. Bak, C. Tang and K. Weisenfeld, *Phys. Rev. Lett.* **59**, 381 (1987).
13. My favorite routes into the fracture literature are *via* B. Lawn, *Fracture of Brittle Solids* (Cambridge University Press, New York, 1993) or L.B. Freund, *Dynamic Fracture Mechanics* (Cambridge University Press, New York, 1990).
14. C. Scholz, *The Mechanics of Earthquakes and Faulting* (Cambridge University Press, New York, 1990).

15. J. M. Carlson, J. S. Langer, and B. E. Shaw, *Rev. Mod. Phys.* **66**, 657 (1994).
16. J. Fineberg, S.P. Gross, M. Marder and H. Swinney, *Phys. Rev. Lett.* **67**, 457 (1991); *Phys. Rev.* **B 45**, 5146 (1992).
17. S.P. Gross, J. Fineberg, M. Marder, W.D. McCormick and H. Swinney, *Phys. Rev. Lett.* **71**, 3162 (1993).
18. E.S.C. Ching, J.S. Langer and Hiizu Nakanishi, *Phys. Rev. Lett.* **76**, 1087 (1996); *Phys. Rev.* **E 53**, 2864 (1996).

# NEW HORIZON IN NUCLEAR PHYSICS AND ASTROPHYSICS USING RADIOACTIVE NUCLEAR BEAMS

Isao TANIHATA

*RIKEN, 2-1 Hirosawa, Wako, Saitama 351-01, Japan*

Beams of $\beta$ - radioactive nuclei, having a lifetime as short as 1 ms have been used for studies of the nuclear structure and reaction relevant to nucleosynthesis in the universe. In nuclear-structure studies, decoupling of the proton and neutron distributions in nuclei has been discovered. The decoupling appeared as neutron halos and neutron skins on the surface of neutron-rich unstable nuclei. In astrophysics, reaction cross sections have been determined for many key reactions of nucleosynthesis involving short-lived nuclei in the initial and final states. One such important reaction, $^{13}$N+p -> $^{14}$O +$\gamma$, has been studied using beams of unstable $^{13}$N nuclei. Such studies became possible after the invention of beams of radioactive nuclei in the mid-80's. Before that, the available ion beams were restricted to ions of stable nuclei for obvious reasons. In the next section the production method of radioactive beams is presented, then a few selected studies using radioactive beams are discussed in the following sections. In the last section, some useful properties of radioactive nuclei for other applications is shown.

## 1. Radioactive Nuclear Beams. -A New Tool -

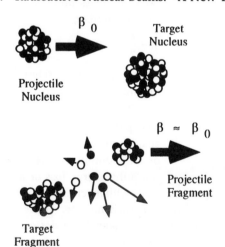

$\beta_0$

Target Nucleus

Projectile Nucleus

$\beta \approx \beta_0$

Projectile Fragment

Target Fragment

Fig. 1 Projectile fragmentation of high-energy heavy-ion collisions. This reaction is used to produce secondary radioactive nuclear beams.

High-energy beams of radioactive nuclei have been produced by the fragmentation of high-energy heavy ions. Projectile fragmentation was discovered by Heckmann and Greiner in Berkeley in the 1970s.[1] When heavy ions of energy up to 2000 MeV per nucleon were incident on target nuclei, they observed that many different kinds of nuclei, many of them being unstable, were emitted into a small forward cone. This phenomenon were understood as a sudden breakup of the projectile nucleus, as illustrated in Fig. 1.

A nucleus is an aggregate of nucleons, protons and neutrons, with binding energy per nucleon of 8 MeV. When these aggregates are smashed against each other with an energy much higher than the binding energy of individual nucleons, the bindings between the nucleons are easily broken, and that part of the projectile not overlapping with the target fly away without any significant perturbation. Different nuclides are produced depending on the number of protons and neutrons remaining in the projectile fragment. The most important property of the projectile fragments is that they come out with the same velocity as that of the incident nuclei. This property provides an easy method for producing usable radioactive beams. [2]

Figure 2 shows one of the radioactive beam separator at RIKEN, the RIPS. Heavy ions of energy up to 135 MeV per nucleon are incident on a production target made of a metallic Be plate. Projectile fragments are guided to the first focus (F1) through a set of magnets, and focused at a different place, depending on their magnetic rigidities. The magnetic rigidity (R) is the ratio of the total momentum of a nucleus and the charge (P/Z).

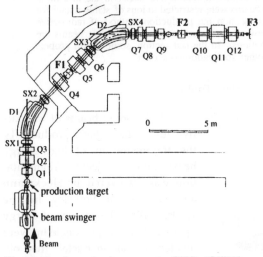

Fig. 2 Radioactive nuclear beam separator (RIPS) at RIKEN.

Because all fragments have the same velocity (β=$v$/c), the separation with the rigidity is equivalent to the separation in $A/Z$, the ratio of the mass number to the proton number,

$$R = \frac{P}{Z} = \frac{A}{Z} \beta \gamma m, \quad (1)$$

where $\gamma$ is the relativistic variable, $\gamma = 1/\sqrt{1-\beta^2}$.

Then, $A/Z$- selected nuclei pass through a wedge-shaped energy degrader made of an aluminum plate. Because the energy loss of a nucleus with a certain velocity is proportional to $Z^2$, the rigidity becomes different for each nuclide. Thus, nuclides are positionally separated at F2 after another set of magnets. A desired nuclide is then selected by a narrow slit at F2, and is transported to F3 where reactions can be studied. This production technique of radioactive nuclear beams is called **the secondary beam method.**

The secondary beam method has a very short separation time. This is because projectile fragments have a velocity faster than 30% of the light speed. Any nuclide that has a lifetime longer than 100 ns can be separated and used as a beam. These types of facilities are now commonly used all over the world. Examples are GANIL in France, GSI in Germany, NSCL at MSU in the USA, and RIKEN in Japan.[3]

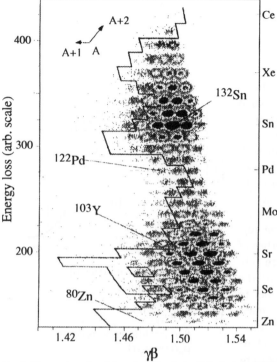

Fig. 3 Observation of nuclei produced by the in-flight fission of $^{238}$U. More than 100 new isotopes have been discovered.

On the other hand, it is difficult to produce a useful low-energy beam by this method. Producing low-energy radioactive nuclear beams requires a different technique. **The reacceleration method** is presently being used. In this method, produced nuclei, by any type of reaction, are put into an ion source, and then accelerated by a set of accelerators. The advantage of this method is the good quality of the beam. Many laboratories around the world are developing efficient and fast methods to produce ions of radioactive nuclei. However, until now acceleration have been possible only for radioactive nuclei with a lifetime longer than several seconds. At this moment, only one facility at Louvain-la-Nouve in Belgium is in full operation. The HIRBF facility at Oak Ridge National Laboratory in the USA will start its operation soon.

The secondary beam method, itself, is also an efficient method for a new isotope search. Using this method, many nuclei have been discovered. Before this method, the neutron drip line ( the limit of neutron-rich nuclei to which no more neutrons can be added to make bound nuclei ) was known only up to Be isotopes. Presently, using this method, a neutron drip line has been obtained up to F isotopes. Also, proton drip line has been attained up to Ge or As isotopes.

This method has recently also been applied for the fission of U. At GSI, $^{238}$U were accelerated up to 1 GeV per nucleon, and shot to a Pb target.[4] Figure 3 shows an identification spectrum of fission products observed at the FRS radioactive beam separator. Because no chemical separation is necessary, all kinds of element have been observed with the same efficiency. In this single experiment, they discovered over one hundred new isotopes. It really shows the power of the method.

## 2. Halos and Skins in Nuclei

### 2.1 Radii of Unstable Nuclei

The radii of stable nuclei have been studied by many methods, such as electron scattering, the spectroscopy of ordinary electron and muonic atoms, and the proton scattering. Those studies have showed that the density distribution can be well described by a smooth function with a flat distribution in most regions and a quick decrease at the surface. Three pronounced properties of the density distribution were observed, as shown in Fig. 4 : (I) the half-density radius is proportional to the one-third power of the mass number $(A)$ ; (ii) the surface diffuseness does not change from nucleus to nucleus ; and (iii) the distributions of protons and neutrons are proportional at all distances even if $Z$ is not

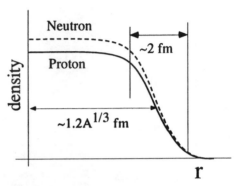

Three common properties of **stable** nuclei

1. $R = r_0 A^{1/3}$
2. Surface diffuseness is constant
3. $\rho_p(r) \propto \rho_n(r)$

Fig. 4 Common properties of stable nuclei.

equal to $N$. These properties immediately draw the saturation property of nuclear matter. Also, because of (iii), one considered that it was impossible to study matter made of only neutrons or protons in the laboratory.

After measuring nuclear radii and the density distribution of unstable nuclei using radioactive nuclear beams, it has been found that these properties are not general, but only common among stable nuclei. A break of these properties has been discovered as neutron skins and neutron halos. We now discuss several examples of such observations in the following.

The radii of unstable nuclei were determined by measuring the interaction cross sections between unstable nuclei and stable nuclei. The interaction cross section is the total probability of removing a nucleon or nucleons from the incident nuclei. Interactions occurs only when a projectile and a target nuclei overlap with each other. Therefore, the interaction-cross section $(\sigma_I)$ is proportional to the overlapping probability,

$$\sigma_I = \pi(R_I + R_2)^2, \qquad\qquad (2)$$

where $R_1$ and $R_2$ are the interaction radii of incident and target nuclei. If we determine $R_2$ based on the collision between identical nuclei, we can then obtain $R_1$ of an incident unstable nucleus. Figure 5 shows thus determined interaction radii of Li, Be, and B isotopes. Although the radii of stable nuclei follow the $A^{1/3}$ rule, neutron-rich nuclei show large deviations.

Not only the interaction radii, but also the root-mean-square (rms) radii, have been determined using models of the interaction-cross sections. The root-mean-square radius ($R_{rms}$) is commonly used for a nucleus and is defined as

$$R^2_{rms} = \frac{4\pi \int r^2 \rho(r) \cdot r^2 dr}{4\pi \int \rho(r) \cdot r^2 dr}, \qquad (3)$$

where $\rho(r)$ is the nucleon-density distribution. $R_{rms}$ has been determined for light nuclei, and is shown in Fig.6.

All metter radii are determined from interaction cross-section measurements at high energy ($\approx$ 800 MeV per nucleon).

In the figure, to easily see the change of radius, the difference in the rms radii from that of $^4$He (the nucleus with the small-

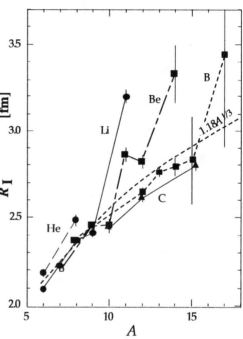

Fig. 5 Interaction radii of light nuclei. It can be seen that some of the nuclei have extremely large radii.

est radius) is presented. The diameter of the circle in the figure shows these differences. This figure clearly shows new properties of nuclear radii. First, as also shown in Fig.5, the radii are not proportional to $A^{1/3}$. In this figure it can be seen as the change of radii along isobars, a group of nuclei having same mass number ($A$.). Nuclei with a larger difference between $N$ and $Z$, or larger isospin nuclei, show larger radii. [$(N-Z)/2$ is called the isospin.] It shows that the radii depend not only on $A_,$, but also on $N$ and $Z$. Second, nuclei at the neutron drip line have extremely large radii. A detailed study of these nuclei has shown a break down of all three common properties observed in stable nuclei, as shown in the following.

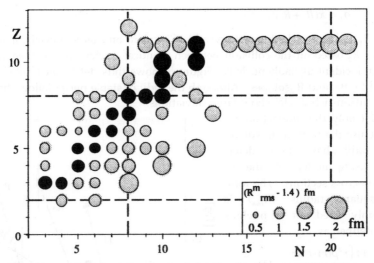

Fig. 6 Matter radii of all p shells and some of sd shell nuclei. All matter radii are determined from interaction cross-section measurements at high energy (≈ 800A MeV).

## 2.2 Neutron Skin

For a stable nucleus, such as $^{208}$Pb or $^{48}$Ca, excess neutrons were expected to form a layer of neutrons on the nuclear surface based on several considerations. However, detailed studies of the proton and neutron distributions have shown no evidence of a surface-neutron excess in any stable nuclei. The rms radii of protons and neutrons were always equal. However, it is now considered that the neutron skin is commonly formed on the surface of unstable nuclei.

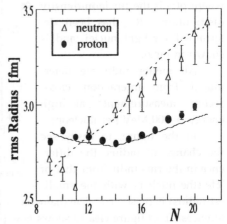

Fig. 7 Proton and neutron radii of Na isotopes.

Figure 7 shows the rms radii of the proton and the neutron distributions in Na isotopes.[5] The proton radii, or the charge radii, of Na isotopes were determined by the laser spectroscopy of neutral atoms of Na isotopes at CERN/ISOLDE in the 70s.[6] The matter radii were recently determined[5] by the radioactive-beam method discussed in the preceding section. Neutron rms radii are then determined from the relation,

$$R_{rms}^{m\,2} = R_{rms}^{p\,2} + R_{rms}^{n\,2} \ . \tag{4}$$

where superscripts $m, p,$ and $n$ indicate a nucleon, proton, and neutron, respectively.

It can be clearly seen that the neutron radius increases much faster than the proton radius for neutron-rich nuclei. This thus clearly shows that neutron skins of a thickness 0.3 to 0.4 fm are formed in neutron-rich Na isotopes. This data are also consistent with previous knowledge, because the proton and neutron radii are equal for the stable isotope $^{23}$Na ($Z$=11, $N$=12). Although it is not decisive from this figure, the formation of proton skins is also suggested for proton-rich nuclei. It obviously shows that rule (iii) is broken for unstable nuclei.

### 2.3   Neutron Halo[7]

Gradual increases in the rms radii for larger isospin nuclei can be explained as the growth of a neutron skin as well as a decrease in the central density. However, sudden large increases in the radius at the drip line is much too large to be explained by the neutron skin. It can be understood by the formation of a neutron halo. The neutron halo is a long tail of the neutron-density distribution extended very far from the center of the nucleus, as shown in Fig. 8. It is now considered that the extremely small binding energy of the last neutron is a cause of the neutron halo.

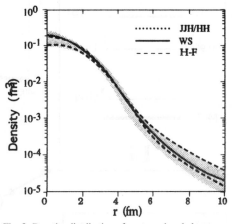

Fig. 8  Denstity distribution of neutrons in a halo nucleus, $^{11}$Li. Very low-density (1/100 of nuclear matter density) neutrons are extended out to a large distance.

Figure 9 illustrates an example for the $^{11}$Li nucleus. In $^{11}$Li the last two neutrons are very weakly bound. The separation energy of the last two neutrons is only 320 keV with a great contrast with the usual nucleon separation energy of 8 MeV. [Fig. 9-a]

60

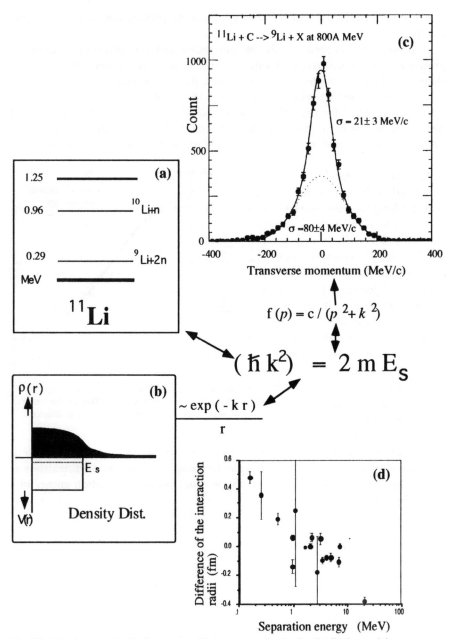

Fig. 9 Relation between the binding energy of halo neutrons, the density distribution, and the momentum distribution.

Let us consider the behavior of such a loosely bound system under a simple potential model. We first treat two neutrons as a single object moving in a square-well potential with a separation energy of $E_s$. The asymptotic form of the s-wave function, $\psi(r)$, of such a system is written by Yukawa function as

$$\psi(r) \propto \frac{\exp(-\kappa r)}{r}, \tag{5}$$

where $\kappa$ is related to the separation energy, as

$$\hbar\kappa^2 = 2\mu E_s , \tag{6}$$

here, $\mu$ is the reduced mass of the system. The smaller the $E_s$, the longer the tail of the distribution. In fact, as shown in the Fig. 9-d, an enhancement of the radii was observed for nuclei with a smaller $E_s$.

On the other hand, the momentum distribution $[f(p)]$ of two neutrons (or the momentum distribution of the core, $^9$Li in this case) is described by,

$$f(p) = \frac{c}{p^2 + \kappa^2} . \tag{7}$$

As can be seen in Fig. 9-c, an extremely narrow distribution was observed in a $^9$Li fragment from $^{11}$Li. This width is about a quarter of that observed in stable nuclei.

Therefore, all observations, the radii, the momentum distribution, and the separation energy, are consistent with each other to show a long tail in the density distribution. Also, experiments with different targets indicated the density distribution shown in Fig. 8.

The observation of a neutron halo showed that the surface diffuseness is not necessarily the same for all nuclei, but depends on the separation energy of the nucleons. In a stable nuclei, the separation energy of the last nucleon is always in between 6-8 MeV. Therefore, surface diffuseness do not change significantly. However, for unstable nuclei, since the separation energy changes widely, the surface diffuseness varies accordingly. The neutron halo is an extreme case of such a weakly bound nucleus. In fact, neutron halos have been observed in many dripline nuclei, such as $^8$He, $^{11}$Li, $^{14}$Be, $^{17, 19}$B, and $^{19}$C. Several candidates of proton halo, such nuclei as $^8$B, $^{17}$F, and $^{17}$Ne, are being investigated. However, the proton halo has not been unambiguously established.

Summarizing this section, the formation of neutron skins and halos has indicated that the three "common" properties of the nuclear-density distribution are not at all common in stable nuclei. (I) The nuclear radius depends not only on $A$, but also $Z$ and $N$. Also, there is a possible indication that the central density of a nucleus depends on $Z$ and $N$. (II) Since the surface diffuseness of a nucleus depends on the separation energy of the nucleon, it thus varies widely for unstable nuclei. (III) The proton and the neutron distributions are decoupled in unstable nuclei.

Therefore, the possibility to study a neutron matter has now been opened.

## 3. New Structure Problems in Nuclei

Studies of unstable nuclei have not only showed a change in the density distribution, but have also opened new problems concerning structure models. Three selected problems are presenting in the following.

### 3.1 Borrowmean System

Near the drip lines, many nuclei show an interesting three-body character. Halo nuclei are often such a case. Let us take $^6$He as an example. In this nucleus, two neutrons are considered to form a neutron halo outside the core of $^4$He. As is well known, two neutrons do not make a bound state. The $^5$He system, $^4$He and a neutron, also do not form a bound state. A $^4$He and two neutrons form a bound-state $^6$He only when three of them are put together. Such a system can be seen in the classical ring combination shown in Fig. 10. This system is called "Borromean rings", since it was used as the family mark of the Italian Borromean family. Similar marks are commonly used also for Kamon, the Japanese family marks.

Efimov predicted that an effective potential for a neutron can have $1/r$ - type dependence if certain conditions are fulfilled.[8] It is extremely interesting if such a situation occurs in a nucleus, because under such a condition infinite numbers of bound level are formed. A huge nucleus may be formed in analogy to the Redberg atom. Nuclear physicists are looking for a precursor of this phenomenon in halo nuclei.

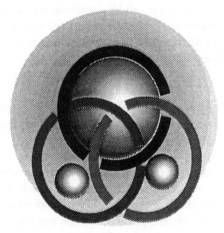

Fig. 10 Borromean rings. Althogh none of the two rings are interlocked, three rings are interlocked together.

### 3.2 Unbalanced Nuclear System

Unstable nuclei are metastable from a wider view point. Although they are bound under strong interactions, they decay by the weak interactions. This situation can be clearly seen in a potential model view, as shown in Fig. 11. In stable nuclei, the one body potentials are the same for protons and neutrons, except for the Coulomb interactions, as shown in Fig. 11-a. Protons and neutrons are then filled in the potential up to the same level. If protons or neutrons are filled to higher than the other, these extra nucleons β decay into the other. Therefore, they are unstable nuclei.

Because of the self-consistency, the shape of the potentials of proton and neutrons are similar in the density distribution. Therefore, the density distributions of protons and neutrons are the same in a stable nucleus. In such a nucleus, most of the nucleons are symmetrically bound and saturated, and many properties of the nucleus are determined by the last few valence nucleons in an unoccupied orbital. The shell model is very effective in such a case.

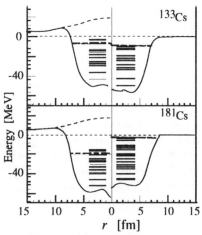

Fig. 11 Potential and single-particle levels for stable and neutron-rich unstable nuclei by a Skyrme Hartree-Fock calculation.

On the other hand, protons and neutrons are filled to different levels in unstable nuclei. In particular, neutrons are filled almost to the top near to the drip line, and protons are deeply bound in contrast. The neutron skins and halos are formed under such a condition. The interactions are not saturated in many nucleons. In contrast to unstable nuclei near the stability line, many nucleons can be involved in the β decay of nuclei near the drip line.

### 3.3 Disappearance and Change of Magic Numbers

The magic numbers, (2, 8, 20, 28, 50, 82) determine the basic properties of nuclei. Extraction of the magic numbers was one of the most significant discoveries in nuclear theory. The magic numbers provide a base for an independent particle and orbital motion picture of a nucleus, which explains many properties of a nucleus.

However, recent observations of neutron-rich nuclei have showed a disappearance of magic numbers. One is the magic number $N=20$. Recently, the strength of the E2 transition between the first excited ($2^+$) state and the ground state has been determined by the Coulomb excitation of the $^{32}$Mg ($Z =12$, $N=20$) nucleus.[9] As can been seen in Fig.12, the E2 transition strength, B(E2), of $^{32}$Mg is much larger than those of the surrounding nuclei. In general, B(E2) is very small for a magic nucleus, because of its spherical nature. The observed large B(E2) indicates that, since $^{32}$Mg is largely deformed, magicity does not exist.

64

Other examples are [10]He and [28]O. Although they have double magic numbers ([10]He:Z=2, N=8, [28]O: Z=8, N=20) and thus were expected to be well bound, recent experiments have shown that [10]He is unbound [10], and bound [28]O [11] has not been observed, and is thus considered to be unbound. Therefore, they do not have as strong a binding energy as expected from the magicity.

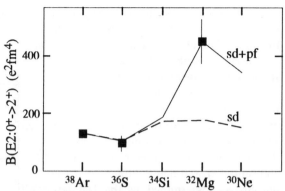

Fig. 12 Strength, B(E2), of the electric-quadrupole transition in N=20 nuclei. The B(E2) of [32]Mg is much larger than those of other nuclei as well as a simple shell model calculation assuming a spherical shape. It thus indicates that [32]Mg is largely deformed and that N=20 is no longer a magic number.

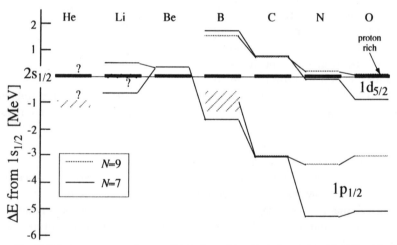

Fig. 13 Relative movement of p, s, and d single-particle orbitals for nuclei with N=7 and 9. Normally well-separated p and s orbitals become closer to each other when protons are removed.

In addition, recent studies of halo nuclei and nuclei near N=8 shows peculiar behavior in single particle levels. Figure 13 shows the relative energy of the $p_{1/2}$, $d_{5/2}$, and $s_{1/2}$ single-particle orbitals. In stable nuclei the order of the orbital is p, d, s, and the energy gap between p and d is large, thus forming the magic number at 8. This can be seen in [15]O and [14]N in the figure. However, the relative energy of these orbital changes when the proton number decreases. The distance between the p and s orbitals becomes smaller, and is even reversed in the case of $^{11}$Be. Also, although the relative energy has not been determined experimentally, strong mixing of $p_{1/2}$ and $s_{1/2}$ is suggested in $^{11}$Li, indicating a closeness of these orbitals.

It is interesting to compare the ordering of the single-particle orbital between neutron-rich nuclei and that of the Coulomb potential (Fig. 14).

The nuclear potential is usually described as a Woods-Saxon-type function having a flat distribution in the central part, and a fast fall off on the surface. The order of the orbitals is as shown in Fig. 14 with the important $l•s$ potential. The magic numbers are also shown.

The order of the orbital in the Coulomb potential is shown on the right-hand side of the figure. The characteristic feature of the Coulomb potential is the mixing of parity. The s and p orbitals degenerate, as can be seen in the n=2 orbital. The ordering of the single-particle levels near $^{11}$Be and $^{11}$Li (typical halo nuclei) are shown in the middle. The closeness of the $s_{1/2}$ and $p_{1/2}$ orbital is similar to that of Coulomb potential. The $l•s$ splitted $p_{3/2}$ orbital is located near the normal energy. This therefore indicates that the order of the orbital is strongly

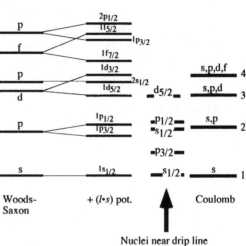

Fig. 14 Observed relative energy of single-particle orbitals around a halo nucleus of $^{11}$Li shows a similarity to those levels in the Coulomb potential.

modified in a weakly bound orbital, and that the new order indicates that the potential has a long tail similar to that of the Coulomb potential $(1/r)$.

The above-mentioned examples indicate that the magic numbers as well as the order of the shell orbital are strongly modified in neutron-rich nuclei, particularly for loosely bound nuclei. In addition, these loosely bound nuclei often show a clustering feature. A consistent theoretical description of these nuclei, including the mean field, three body and fundamental interactions, and the clustering features is required.

## 4. Astrophysical and Other Applications

Radioactive nuclear beams are also applied in other fields of study. Here only a few examples are presented as an introduction. Any interested reader is referred to read the articles shown in the list of references.[12]

### 4.1 Nucleosynthesis in hot stars and supernovae

#### Rapid proton burning

A number of extreme astrophysical environments have the capacity to undergo rapid proton burning, known as the rp process. One such environment is the region surrounding a collapsed star - a white dwarf or a neutron star - in a binary system, where the collapsed star accretes matter from its companion. Such environments would have temperatures of several hundred million K, where the times between successive proton radiative capture reactions ( such as $^{12}C + p \rightarrow ^{13}N + \gamma$ ) would become much shorter than those characteristic of $\beta$ decays. The $\beta$ decays ordinarily intercede after roughly every other proton radiative capture, thus maintaining the rough equivalence between neutron and proton number characteristic of stable nuclei. A proton-burning scenario is sketched in Fig.15, which shows both the usual CNO cycle, by which C catalyzes the conversion of four protons (with two $\beta$ decays ) into a $^4He$ nucleus, and the hot CNO, or HCNO, cycle.

Although $\beta$ decay from $^{13}N$ to $^{13}C$ always occurs before the next proton capture at normal stellar temperatures of the CNO cycle. On the other hand in a high-temperature environment proton-capture probability becomes higher, then $^{14}O$ can be made via the $^{13}N + p \rightarrow ^{14}O + \gamma$ reaction. This reaction has therefore been an archetypal one in radioactive nuclear-beam research.

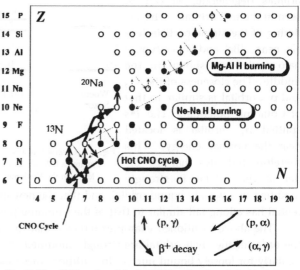

Fig. 15 The CNO cycle and the hot CNO cycle. A cycle that includes heavier nuclei is realized when (p,γ) reactions of unstable nuclei become faster than b decay.

The rate of the $^{13}N + p \rightarrow ^{14}O + \gamma$ reaction has been measured by several methods using radioactive beam of $^{13}N$ or $^{14}O$. One measurement, studied the reaction using the $^{13}N$-beam facility at Louvain-la-Neuve. [13] Two other measurements were based on a recently developed technique, known as Coulomb breakup,[14] which involves measuring the cross section through an inverse reaction. This approach requires an $^{14}$O beam, available only at a radioactive nuclear beam facility. A heavy nucleus, for example lead, Coulomb - excites the $^{14}$O nuclei to states that decay by proton emission. Then, through coincidence detection, they measured the cross section for producing a proton and a $^{13}$N residual nucleus. This approach holds great promise for measuring a number of reactions that cannot otherwise be studied, such as those involving neutron capture on a short-lived nucleus.

The Coulomb-excitation method is now being widely used to study (p, $\gamma$), (n, $\gamma$), (d, $\gamma$) reactions relevant to astrophysics. One such reaction is the $^7$Be(p, $\gamma$)$^8$B reaction, which is of key importance concerning the solar-neutrino problem.

Reactions involving somewhat heavier nuclides than those shown in Fig. 14 require somewhat higher temperatures simply to overcome the Coulomb barriers. Furthermore, rapid hydrogen burning of the more massive nuclides can be extremely complex to describe, because one must know the cross sections of the reactions (mostly of proton radiative captures) that couple all of the relevant nuclides. Recent studies of the most proton-rich nuclides that are stable to nucleon decay have allowed nuclear physicists to determine the limiting masses for those nuclides that must be considered in the rp process. Specifically, the radioactive-nuclear-beam facility at Michigan State University was used to find very proton rich nuclides, such as $^{63}$Ga, $^{62, 63}$Ge, $^{65}$As, $^{69}$Br, $^{75}$Sr, which are stable to nucleon emission; that is, they $\beta$ decay. [15] None of these nuclides had been observed previously. These observation shows that rp-process would continue at least upto those elements.

### Rapid neutron capture

The stellar R process is thought to occur by successive radiative captures of many neutrons on pre-existing "seed nuclei" in the core of a supernova. This process generates nuclides with more than ten more neutrons than the most neutron-rich stable nuclei, thus driving the nuclides to the neutron drip line. For nuclei between nuclear shell closures these neutron captures occur rapidly. However, at shell closures a neutron must $\beta$ decay to a proton before a subsequent neutron capture can occur. Thus, the rate at which the R process can proceed is mediated by the half-lives of these "waiting point" nuclei. The predicted path of the R process is through the neutron rich nuclei where most of them are not even synthesised in the laboratory yet. Only planned new generation facilities will be able to produce these R-process nuclei.

Some of the first experiments with beams of short-lived nuclei, involving nuclides important to the R process, were to determine the half-lives and information about nuclear levels of neutron rich nuclides such as $^{80}$Zn, which has

68

12 neutron more than the most massive stable Zn isotope but, with 50 neutrons, has a closed shell.

It may also be important to study reactions on some of the neutron - rich nuclides important to the R process. Because measuring the neutron radiative capture cross sections would require radioactive-nuclear beams incident on neutron targets, they are not likely to be carried out in the near future. However, the Coulomb-breakup technique could provide important means in many cases. The production of a beam of residual nuclide in the reaction of interest, followed by coincidence detection of the neutron and the heavy reaction-product nucleus resulting from a Coulomb breakup of the beam nuclei, could provide the desired cross sections, provided excited states of the initial- and final-state nuclides do not complicate the picture.

In any case, we need a next-generation radioactive nuclear beam facility to study the R process.

## 4.2 Combination of radioactive nuclear beams and PET

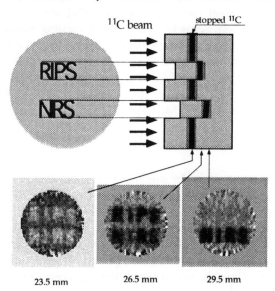

23.5 mm        26.5 mm        29.5 mm

Fig. 16 Combination of Position-Emission-Tomography and radioactive nuclear beams provides a new tool for cancer treatment.

Positron Emission Tomography (PET) is being effectively used in medical diagnoses. On the other hand, heavy ions are being used to treat cancer. The advantage of heavy ions for cancer treatment is due to the sharpness of the dose distribution. Therefore, heavy ions can kill cancer cells with minimal effects to the normal cells that are sitting next to the tumor. To take maximum advantage of this property it is extremely important to locate the cancer position and to know it's shape. In addition, a fine tuning of the amount of dose and distribution is fundamental. One disadvantage of heavy ions is that the irradiated positions, although they are carefully calibrated and simulated before a treatment, are not directly observed during the treatment.

The use of beams of positron-emitting radioisotopes provides a means to monitor beam irradiation. Figure 16 shows such a test experiment at the RIKEN radioactive beam facility under collaboration between RIKEN and NIRS (National Institute of Radiological Sciences).

A beam of $^{11}C$ was injected into a lucite plate. The name of the radioactive beam separator at RIKEN (RIPS) and NIRS are carved on the surface of a plate with a slightly different depth. The thickness of the lines of characters is 1 mm. The implanted $^{11}C$ were sectionaly imaged by PET. As can be seen from the three images shown in the figure, it was demonstrated that the implanted position of $^{11}C$ can be determined with a resolution better than 1 mm. It shows that radioactive beams can be used for cancer treatment providing simultaneous monitoring of the dose distribution.This method also provides a new detection method of an elemental flow or diffusion in a material if PET images are taken within certain time intervals.

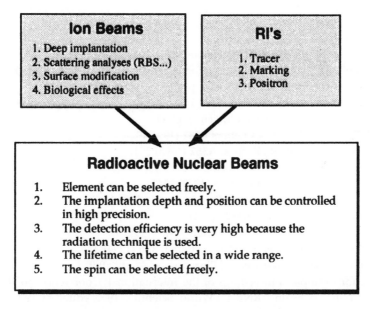

Fig. 17 Good properties of radioactive nuclear beams for applications.

Radioactive nuclear beams have useful properties for many applications. They have combined properties of high-energy heavy-ions and radioactive isotopes. Such useful properties are summarized in Fig. 17.

## Summary

Radioactive beams of low and high energies have recently been developing, and are being used in laboratories all over the world. New discoveries in nuclear structure, such as the neutron halo and the neutron skin, have been made. Those new structures call for the necessity of more unified models in nuclear physics. Several important reaction probabilities for nucleosynthesis at various cosmological stages and environments have been determined for the first time. Because most of the important reactions for understanding nucleosyntheses are those including radioactive nuclei, studies with radioactive nuclear beams are of fundamental importance. Also, radioactive beams are expected to open up new applications in chemistry, solid-state physics, biology, medicine, and others.

With a wide range of expectations, many new facilities are under construction, and are being proposed in many countries. As one future project, RIKEN has proposed the RI Beam Factory. This facility is expected to deliver more than 2000 radioactive nuclei as beams.

References
[1] A.S. Goldhaber and H.H. Heckman, Ann. Rev. Nucl. Part. Sci. (1978).
[2] I. Tanihata, Treatise on Heavy-Ion Science, Vol. 8 (1989), Edited by D.A. Bromley (Plenum Pub. Co.) ch. 6, p443.
[3] I. Tanihata, Progr. in Part. and Nucl. Phys. Vol 35 (1995), Edited by A. Faessler, Elsevier Science Ltd., ch. 6, p505.
[4] M. Bernas et al., Physics Letters B331 (1994)19.
    Ch. Engelmann et al., Z Phys. A352 (1995) 351.
[5] T. Suzuki et al., Phys. Rev. Letters 75 (1995) 3241.
[6] G. Huber et al., Phys. Rev. C18 (1978) 2342.
[7] I. Tanihata, J. Phys. G : Nucl. Part. Phys. 22 ( 1996 ) 157.
[8] V. Efimov, Nucl. Phys. A362 (1981) 45.
[9] T. Motobayashi et al., Phys. Letters B346 (1995) 9.
[10] A. Korcheninnikov et al., Phys. Letters B 326 (1994) 31.
[11] D. Guillemand-Mueller et al., Phys. Rev. C 41 (1990) 937.
[12] See for example, Proceedings of the International Conference on Radioactive Nuclear Beams, RNB-1 ; World Scientific 1990, edited by W.D. Myers, J.M. Nitschke, and E.B. Norman,
    RNB-2 ; Adam Hilgar Pub.Co. 1992, edited by Th. Delbar, RNB-3 ; Editions Frontiers 1994, edited by D.J. Morrissey.
    RNB-3; Proceedings of the Third International Conference on Radioactive Nuclear Beams, 24-27 May, 1993 MSU East Lansing, Michigan, USA edited by D. J. Morrissey.
[13] P. Decrock et al., Phys. Rev. Letters 67 (1991) 808.
[14] T. Motobayashi et al., Phys. Letters B 264 (1991) 259.
[15] M.F. Mohar et al., Phys Rev. Letters 66 (1991) 1571.

# THE FUTURE OF PARTICLE PHYSICS

FRANK WILCZEK

*School of Natural Sciences, Institute for Advanced Study*
*Princeton, NJ 08540, USA*

In the first part of the talk, I give a low-resolution overview of the current state of particle physics – the triumph of the Standard Model, and its discontents. I review and re-endorse the remarkably direct and (to me) compelling argument that existing data, properly interpreted, point toward a unified theory of fundamental particle interactions and toward low-energy supersymmetry as a major component of the near-term future of high energy physics. I then briefly discuss other challenges and opportunities in the field.

## 1 Triumph of the Standard Model

The core of the Standard Model [1,2,3] of particle physics is easily displayed in a single Figure, here Figure 1. There are gauge groups $SU(3) \times SU(2) \times U(1)$ for the strong, weak, and electromagnetic interactions. The gauge bosons associated with these groups are minimally coupled to quarks and leptons according to the scheme depicted in the Figure. The non-abelian gauge bosons within each of the $SU(3)$ and $SU(2)$ factors also couple, in a canonical minimal form, to one another. The $SU(2) \times U(1)$ group is spontaneously broken to the $U(1)$ of electromagnetism. This breaking is parameterized in a simple and (so far) phenomenologically adequate way by including an $SU(3) \times SU(2) \times U(1)$ $(1, 2, -\frac{1}{2})$ scalar 'Higgs' field which condenses, that is, has a non-vanishing expectation value in the ground state. Condensation occurs at weak coupling if the bare $(\text{mass})^2$ associated with the Higgs doublet is negative.

The fermions fall into five separate multiplets under $SU(3) \times SU(2) \times U(1)$, as depicted in the Figure. The color $SU(3)$ group acts horizontally; the weak $SU(2)$ vertically, and the hypercharges (equal to the average electric charge) are as indicated. Note that left- and right-handed fermions of a single type generally transform differently. This reflects parity violation. It also implies that fermion masses, which of course connect the left- and right-handed components, only arise upon spontaneous $SU(2) \times U(1)$ breaking.

Only one fermion family has been depicted in Figure 1; of course in reality there are three repetitions of this scheme. Also not represented are all the complications associated with the masses and Cabibbo-like mixing angles among the fermions. These masses and mixing angles are naturally accommodated as parameters within the Standard Model, but I think it is fair to say that they are not much related to its core ideas – more on this below.

$$SU(3) \quad \times \quad SU(2) \quad \times \quad U(1)$$

$$\text{8 gluons} \qquad W^{\pm}, Z \qquad \gamma$$

$$\underline{\qquad\qquad}$$

mixed

$$SU(3)$$

$$\longleftrightarrow$$

$$SU(2) \updownarrow \begin{pmatrix} u_L^r & u_L^w & u_L^b \\ d_L^r & d_L^w & d_L^b \end{pmatrix} \frac{1}{6} \qquad (u_R^r \quad u_R^w \quad u_R^b) \frac{2}{3}$$
$$(d_R^r \quad d_R^w \quad d_R^b) - \frac{1}{3}$$

$$\begin{pmatrix} \nu_L \\ e_L \end{pmatrix} - \frac{1}{2} \qquad e_R - 1$$

Figure 1: The core of the Standard Model: the gauge groups and the quantum numbers of quarks and leptons. There are three gauge groups, and five separate fermion multiplets (one of which, $e_R$, is a singlet). Implicit in this Figure are the universal gauge couplings – exchanges of vector bosons – responsible for the classic phenomenology of the strong, weak, and electromagnetic interactions. The triadic replication of quark and leptons, and the Higgs field whose couplings and condensation are responsible for $SU(2) \times U(1)$ breaking and for fermion masses and mixings, are not indicated.

With all these implicit understandings and discrete choices, the core of the Standard Model is specified by three numbers – the universal strengths of the strong, weak, and electromagnetic interactions. The electromagnetic sector, QED, has been established as an extraordinarily accurate and fruitful theory for several decades now. Let me now briefly describe the current status of the remainder of the Standard Model.

Some recent stringent tests of the electroweak sector of the Standard Model are summarized in Figure 2. In general each entry represents a very different experimental arrangement, and is meant to test a different fundamental aspect of the theory, as described in the caption. There is precisely one parameter (the mixing angle) available within the theory, to describe all these measurements. As you can see, the comparisons are generally at the level of a per cent accuracy or so. Overall, the agreement appears remarkably good, especially to anyone

| | | Measurement with Total Error | Systematic Error | Standard Model | Pull |
|---|---|---|---|---|---|
| | $\alpha(m_Z^2)^{-1}$ | $128.896 \pm 0.090$ | 0.083 | 128.907 | $-0.1$ |
| a) | **LEP** | | | | |
| | line-shape and lepton asymmetries: | | | | |
| | $m_Z$ [GeV] | $91.1863 \pm 0.0020$ | 0.0015 | 91.1861 | 0.1 |
| | $\Gamma_Z$ [GeV] | $2.4946 \pm 0.0027$ | 0.0017 | 2.4960 | $-0.5$ |
| | $\sigma_h^0$ [nb] | $41.508 \pm 0.056$ | 0.055 | 41.465 | 0.8 |
| | $R_\ell$ | $20.778 \pm 0.029$ | 0.024 | 20.757 | 0.7 |
| | $A_{FB}^{0,\ell}$ | $0.0174 \pm 0.0010$ | 0.007 | 0.0159 | 1.4 |
| | + correlation matrix | | | | |
| | $\tau$ polarisation: | | | | |
| | $\mathcal{A}_\tau$ | $0.1401 \pm 0.0067$ | 0.0045 | 0.1458 | $-0.9$ |
| | $\mathcal{A}_e$ | $0.1382 \pm 0.0076$ | 0.0021 | 0.1458 | $-1.0$ |
| | b and c quark results: | | | | |
| | $R_b^0$ | $0.2179 \pm 0.0012$ | 0.0009 | 0.2158 | 1.8 |
| | $R_c^0$ | $0.1715 \pm 0.0056$ | 0.0042 | 0.1723 | $-0.1$ |
| | $A_{FB}^{0,b}$ | $0.0979 \pm 0.0023$ | 0.0010 | 0.1022 | $-1.8$ |
| | $A_{FB}^{0,c}$ | $0.0733 \pm 0.0049$ | 0.0026 | 0.0730 | 0.1 |
| | + correlation matrix | | | | |
| | $q\bar{q}$ charge asymmetry: | | | | |
| | $\sin^2\theta_{\text{eff}}^{\text{lept}}$ ($\langle Q_{FB}\rangle$) | $0.2320 \pm 0.0010$ | 0.0008 | 0.23167 | 0.3 |
| b) | **SLD** | | | | |
| | $\sin^2\theta_{\text{eff}}^{\text{lept}}$ ($A_{LR}$) | $0.23061 \pm 0.00047$ | 0.00014 | 0.23167 | $-2.2$ |
| | $R_b^0$ | $0.2149 \pm 0.0038$ | 0.0021 | 0.2158 | $-0.2$ |
| | $\mathcal{A}_b$ | $0.863 \pm 0.049$ | 0.032 | 0.935 | $-1.4$ |
| | $\mathcal{A}_c$ | $0.625 \pm 0.084$ | 0.041 | 0.667 | $-0.5$ |
| c) | $p\bar{p}$ and $\nu$N | | | | |
| | $m_W$ [GeV] ($p\bar{p}$) | $80.356 \pm 0.125$ | 0.110 | 80.353 | 0.0 |
| | $1 - m_W^2/m_Z^2$ ($\nu$N) | $0.2244 \pm 0.0042$ | 0.0036 | 0.2235 | 0.2 |
| | $m_t$ [GeV] ($p\bar{p}$) | $175 \pm 6$ | 4.5 | 172 | 0.5 |

Figure 2: A recent compilation of precision tests of electroweak theory, from [4] , to which you are referred for details. Despite some 'interesting' details, clearly the evidence for electroweak $SU(2) \times U(1)$, with the simplest doublet-mediated symmetry breaking pattern, is overwhelming.

familiar with the history of weak interactions.

Some recent stringent tests of the strong sector of the Standard Model are summarized in Figure 3[5]. Again a wide variety of very different measurements are represented, as indicated in the caption. A central feature of the theory

74

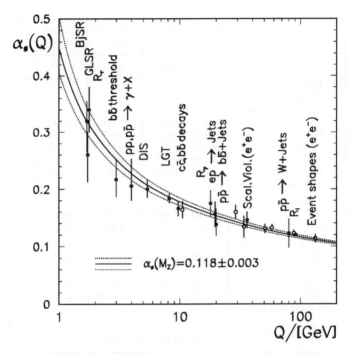

Figure 3: A recent compilation of tests of QCD and asymptotic freedom, from [5], to which you are referred for details. Results are presented in the form of determinations of the effective coupling $\alpha_s(Q)$ as a function of the characteristic typical energy-momentum scale involved in the process being measured. Clearly the evidence for QCD in general, and for the decrease of effective coupling with increasing energy-momentum scale (asymptotic freedom) in particular, is overwhelming.

(QCD) is that the value of the coupling, as measured in different physical processes, will depend in a calculable way upon the characteristic energy scale of the process. The coupling was predicted – and evidently is now convincingly measured – to decrease as the inverse logarithm of the energy scale: asymptotic freedom. Again, all the experimental results must be fit with just one parameter – the coupling at any single scale, usually chosen as $M_Z$. As you can see, the agreement between theory and experiment is remarkably good. The accuracy of the comparisons is at the 1-2 % level.

Let me emphasize that these Figures barely begin to do justice to the evidence for the Standard Model. Several of the results in them summarize quite a large number of independent measurements, any one of which might have falsified the theory. For example, the single point labeled 'DIS' in Figure

3 describes literally hundreds of measurements in deep inelastic scattering with different projectiles and targets and at various energies and angles, which must - – if the theory is correct – all fit into a tightly constrained pattern.

I last reviewed this situation on a related occasion several months ago. At that time, there were reported discrepancies between experimental observations and the Standard Model prediction of the branching ratio $R_b$ of the $Z$ into $b$ quarks, and also with the Standard Model (QCD) prediction of inclusive jet production at high transverse energy. In the meantime these discrepancies have come to seem much less significant: for $R_b$, mostly because of the inclusion of new data; for the jet production, because of a better appreciation of the uncertainties in existing structure function parametrizations. Thus once (or rather twice) again, the Standard Model has survived the challenges that inevitably accompany stiff scrutiny. Another small but long-standing and annoying anomaly, the slightly high value of the strong coupling $\alpha_s(M_Z)$ inferred from the width of the $Z$ has also disappeared – largely, I am told, because the effect of passing trains perturbing the beam energy and thus causing a spurious 'widening' has now been accounted for!

The central theoretical principles of the Standard Model have been in place for nearly twenty-five years. Over this interval the quality of the relevant experimental data has become incomparably better, yet no essential modifications of these venerable principles has been required. Let us now praise the Standard Model:

• The Standard Model is here to stay, and describes *a lot*.

Since there is quite direct evidence for each of its fundamental ingredients (*i.e.* its interaction vertices), and since the Standard Model provides an extremely economical packaging of these ingredients, I think it is a safe conjecture that it will be used, for the foreseeable future, as the working description of the phenomena within its domain. And this domain includes a very wide range of phenomena – indeed not only what Dirac called "all of chemistry and most of physics"[a], but also the original problems of radioactivity and nuclear interactions which inspired the birth of particle physics in the 1930s, and much that was unanticipated.

• The Standard Model is a *principled* theory.

Indeed, its structure embodies a few basic principles: special relativity, locality, and quantum mechanics, which lead one to quantum field theories, local symmetry (and, for the electroweak sector, its spontaneous breakdown),

---

[a]Dirac was referring, here, to quantum electrodynamics.

and renormalizability (minimal coupling). The last of these principles, renormalizability, may appear rather technical and perhaps less compelling than the others; we shall shortly have occasion to re-examine it in a larger perspective. In any case, the fact that the Standard Model is principled in this sense is profoundly significant: it means that its predictions are precise and unambiguous, and generally cannot be modified 'a little bit' except in very limited, specific ways. This feature makes the experimental success especially meaningful, since it becomes hard to imagine that the theory could be approximately right without in some sense being exactly right.

• The Standard Model *can be extrapolated.*

Specifically because of the asymptotic freedom property, one can extrapolate using the Standard Model from the observed domain of experience to much larger energies and shorter distances. Indeed, the theory becomes simpler – the fundamental interactions are all effectively weak – in these limits. The whole field of very early Universe cosmology depends on this fact, as do the impressive semi-quantitative indications for unification and supersymmetry I shall be emphasizing momentarily.

## 2   Deficiencies of the Standard Model

Just because it is so comprehensive and successful, we should judge the Standard Model by demanding criteria. It is clearly an important part of the Truth; the interesting question becomes: How big a part? Critical scrutiny reveals several important shortcomings of the Standard Model:

• The Standard Model contains scattered multiplets with peculiar hypercharge assignments.

While little doubt can remain that the Standard Model is essentially correct, a glance at Figure 1 is enough to reveal that it is not a complete or final theory. The fermions fall apart into five lopsided pieces with peculiar hypercharge assignments; this pattern needs to be explained. Also the separate gauge theories, which are quite mathematically and conceptually similar, are fairly begging to be unified.

• The Standard Model supports the possibility of strong P and T violation [6].

There is a near-perfect match between the necessary 'accidental' symmetries of the Standard Model, dictated by its basic principles as enumerated above, and the observed symmetries of the world. The glaring exception is that there is an allowed – gauge invariant, renormalizable – interaction which, if present, would induce significant violation of the discrete symmetries $P$ and $T$ in the strong interaction. This is the notorious $\theta$ term. $\theta$ is an angle which *a priori* one might expect to be of order unity, but in fact is constrained by experimental limits on the neutron electric dipole moment to be $\theta \lesssim 10^{-8}$. This problem can be addressed by postulating a sort of quasi-symmetry (Peccei-Quinn [7] symmetry), which roughly speaking corresponds to promoting $\theta$ to a dynamical variable – a quantum field. The quanta of this field [8], *axions*, provide an interesting dark matter candidate [9]. Other possibilities for explaining the absence of strong $P$ and $T$ violation have been proposed, but they require towers of hypotheses which seem to me quite fragile.

In no way, of course, should the absence of strong $P$ and $T$ violation be taken as evidence against QCD itself. For practical purposes, one can simply take $\theta$ as a parameter to be fixed experimentally. One finds it to be very small – and is done with it!

• The Standard Model does not address family problems.

There are several distinct 'family problems', ranging from the extremely qualitative (digital – why *three* families?) to the semi-qualitative (some distinctive patterns – why does like couple to like, with small mixing angles?) to the straightforward but most challenging goal of doing full justice to experience by computing (analog) experimental numbers with controlled, small fractional errors:

Why are there three repeat families? Rabi's famous question regarding the muon – "Who ordered *that?*" – still has no convincing answer.

How does one explain the very small electron mass? The dimensionless coupling associated with the electron mass, that is its strength of Yukawa coupling to the Higgs field, is about $g_e \sim 2 \times 10^{-6}$. It is almost as small as the limits on the $\theta$ angle (suggesting, perhaps, that strong P and T violation is just around the corner?). This question can of course be generalized – all the fermion masses, with one exception, are sufficiently small to beg qualitative explanation. The exception, of course, is the $t$ quark. It is a fascinating and important possibility that roughly the observed value of $t$ quark mass at low energies might result after running from a wide range of fundamental couplings at a high scale [10]. If so, one would have a satisfactory qualitative explanation of the value of this parameter.

Why do the weak currents couple approximately in the order of masses? That is, light with light, heavy with heavy, and intermediate with intermediate. Why are the mixings what they are – small, but not miniscule? The same, for the CP violating phase in the weak currents (parameterized invariantly by Jarlskog's $J$) [11]? - - and, by the way, are we sure that $\theta \ll J$? And so on ...

• The Standard Model does not allow non-vanishing neutrino masses.

This is the only entry on my list that has a primarily *experimental* motivation. At present there are three quite different experimental hints for non-vanishing neutrino masses: the solar neutrino problem [12], the atmospheric neutrino problem [13], and the Los Alamos oscillation experiment [14]. The Standard Model in its conventional form does not allow non-zero neutrino masses. However I would like to mention that only a very minimal extension of the theory is necessary to accommodate such masses. One can add a complete $SU(3) \times SU(2) \times U(1)$ singlet fermion $N_R$ to the model. $N_R$ can be given, consistent with all symmetries and with the requirement of renormalizability, a Majorana mass $M$. Note especially that such a mass does not violate $SU(2) \times U(1)$. Likewise, $N_R$ can have a Yukawa coupling to the ordinary lepton doublets through the Higgs field. Then condensation of the Higgs field activates the "see-saw" mechanism [15] to give a small mass for the observed neutrinos; with $M \lesssim 10^{15}$ Gev a range of experimentally and perhaps cosmologically interesting neutrino masses can be accommodated.

• Gravity is not included in the Standard Model.

This really represents (at least) two logically separate problems.

First there is the ultraviolet problem, the notorious non-renormalizability of quantum gravity. This provides an appropriate context, in which to introduce the modern perspective toward the whole concept of renormalizability.

Suppose that one were to be naive, and simply add the Einstein Lagrangian for general relativity to the Standard Model, of course coupling the matter fields appropriately (minimally). Following Feynman and many others, one could then derive, formally, the perturbation series for any physical process. One would find, however, that, the integrals over closed loops generally diverges at the high-energy (ultraviolet) end. Indeed the graviton coupling has, in units where $\hbar = c = 1$, dimensions of $M_{\text{Planck}}^{-1}$ [b]. Here $M_{\text{Planck}} \approx 10^{19}$ Gev is a measure of the stiffness of space-time. Thus higher and higher order terms will, on dimensional grounds, introduce higher and higher factors of the ultraviolet cutoff to compensate. However if we determine (by notional experiments) the couplings at a given scale $\Lambda << M_{\text{Planck}}$ and calculate corrections by only including energy-momenta between the scale $P < \Lambda$ of interest and $\Lambda$, the successive terms in perturbation theory will be accompanied by positive powers of $\frac{\Lambda}{M_{\text{Planck}}}$ and will be small. Thus we can, for example, consistently set all non-minimal couplings to zero at any chosen energy-momentum scale well

---

[b]This occurs because the kinetic energy for the graviton arises from the Einstein action $\propto M_{\text{Planck}}^2 \int \sqrt{g} R$ so that in expanding about flat space one must take $g_{\mu\nu} = \eta_{\mu\nu} + \frac{1}{M_{\text{Planck}}} h_{\mu\nu}$ in order to obtain a properly normalized quadratic kinetic term for $h$.

below the Planck scale. They will then be negligibly small for all practically accessible scales. For different choices of the scale they will be different, but since they are negligible in any case that hardly matters. This procedure is, in practice, the one we always adopt – and the Standard Model peacefully coexists with gravity, so long as we refuse to consider $P \gtrsim M_{\text{Planck}}$.

However from this perspective a second problem looms larger than ever. The energy (and negative pressure) density of matter-free space, the notorious cosmological term, occurs as the coefficient $\lambda$ of the identity term in the action: $\delta \mathcal{L} = \lambda \int \sqrt{g}$. It has dimensions of $(\text{mass})^4$, and on phenomenological grounds we must suppose $\lambda \lesssim (10^{-12} \text{ Gev })^4$. The question is: where does such a tiny scale come from? What is so special about the present state of the Universe, that the value of the effective $\lambda$ for it, which one might naively expect to reflect contributions from much higher scales, is so effectively zeroed?

There may also be problems with forming a fully consistent quantum theory even of low-energy processes involving black holes [16].

Finally I will mention a question that I think has a rather different status from the foregoing, although many of my colleagues would put it on the same list, and maybe near the top:

The Standard Model begs the question "*Why* does $SU(2) \times U(1)$ symmetry break?".

To me, this is an example of the sort of metaphysical question that could easily fail to have a meaningful answer. There is absolutely nothing wrong, logically, with the classic implementation of the Higgs sector as described earlier. Nevertheless, one might well hunger for a wider context in which to view the existence of the Higgs doublet and its negative $(\text{mass})^2$ – or a suitable alternative.

## 3   Unification: Symmetry

Each of the deficiencies of the Standard Model mentioned in the previous section has provoked an enormous literature, literally hundreds or thousands of papers. Obviously I cannot begin to do justice to all this work. Here I shall concentrate on the first question, that of deciphering the message of the scattered multiplets and peculiar hypercharges. Among our questions, this one has so far inspired the most concrete and compelling response – a story whose implications range far beyond the question which inspired it.

Given that the strong interactions are governed by transformations among three color charges – say RWB for red, white, and blue – while the weak interactions are governed by transformations between two others – say GP for

80

SU(5): 5 colors RWBGP

$\underline{10}$: 2 different color labels (antisymmetric tensor)

$$
\begin{array}{llll}
u_L : & RP, & WP, & BP \\
d_L : & RG, & WG, & BG \\
u_L^c : & RW, & WB, & BR \\
 & (\bar{B}) & (\bar{R}) & (\bar{W}) \\
e_L^c : & GP & & \\
 & (\ ) & &
\end{array}
\quad
\begin{pmatrix}
0 & u^c & u^c & u & d \\
 & 0 & u^c & u & d \\
 & & 0 & u & d \\
 & * & & 0 & e \\
 & & & & 0
\end{pmatrix}
$$

$\underline{5}$: 1 anticolor label

$$
\begin{array}{ll}
d_L^c : & \bar{R}, \quad \bar{W}, \quad \bar{B} \\
e_L : & \bar{P} \\
\nu_L : & \bar{G}
\end{array}
\qquad (d^c \quad d^c \quad d^c \quad e \quad \nu)
$$

$$\boxed{Y = -\tfrac{1}{3}(R+W+B) + \tfrac{1}{2}(G+P)}$$

Figure 4: Organization of the fermions in one family in $SU(5)$ multiplets. Only two multiplets are required. In passing from this form of displaying the gauge quantum numbers to the form familiar in the Standard Model, one uses the bleaching rules R+W+B = 0 and G+P = 0 for $SU(3)$ and $SU(2)$ color charges (in antisymmetric combinations). Hypercharge quantum numbers are identified using the formula in the box, which reflects that within the larger structure $SU(5)$ one only has the combined bleaching rule R+W+B+G+P = 0. The economy of this Figure, compared to Figure 1, is evident.

green and purple – what could be more natural than to embed both theories into a larger theory of transformations among all five colors? This idea has the additional attraction that an extra U(1) symmetry commuting with the strong SU(3) and weak SU(2) symmetries automatically appears, which we can attempt to identify with the remaining gauge symmetry of the Standard Model, that is hypercharge. For while in the separate SU(3) and SU(2) theories we must throw out the two gauge bosons which couple respectively to the color combinations R+W+B and G+P, in the SU(5) theory we only project out R+W+B+G+P, while the orthogonal combination (R+W+B)-$\frac{3}{2}$(G+P) remains.

Georgi and Glashow [17] originated this line of thought, and showed how it could be used to bring some order to the quark and lepton sector, and in particular to supply a satisfying explanation of the weird hypercharge assignments in the Standard Model. As shown in Figure 4, the five scattered SU(3)×SU(2)×U(1) multiplets get organized into just two representations of $SU(5)$. It is an extremely non-trivial fact that the known fermions fit so smoothly into $SU(5)$ multiplets.

In all the most promising unification schemes, what we ordinarily think of as matter and anti-matter appear on a common footing. Since the fundamental gauge transformations do not alter the chirality of fermions, in order to represent the most general transformation possibilities one should use fields of one chirality, say left, to represent the fermion degrees of freedom. To do this, for a given fermion, may require a charge conjugation operation. Also, of course, once we contemplate changing strong into weak colors it will be difficult to prevent quarks and leptons from appearing together in the same multiplets. Generically, then, one expects that in unified theories it will not be possible to make a global distinction between matter and anti-matter and that both baryon number $B$ and lepton number $L$ will be violated, as they definitely are in $SU(5)$ and its extensions.

As shown in Figure 4, there is one group of ten left-handed fermions that have all possible combinations of one unit of each of two different colors, and another group of five left-handed fermions that each carry just one negative unit of some color. (These are the ten-dimensional antisymmetric tensor and the complex conjugate of the five-dimensional vector representation, commonly referred to as the "five-bar".) What is important for you to take away from this discussion is not so much the precise details of the scheme, but the idea that *the structure of the Standard Model, with the particle assignments gleaned from decades of experimental effort and theoretical interpretation, is perfectly reproduced by a simple abstract set of rules for manipulating symmetrical symbols.* Thus, for example, the object RB in this Figure has just the strong, electromagnetic, and weak interactions we expect of the complex conjugate of the right-handed up-quark, without our having to instruct the theory further. If you've never done it I heartily recommend to you the simple exercise of working out the hypercharges of the objects in Figure 4 and checking against what you need in the Standard Model - – after doing it, you'll find it's impossible ever to look at the standard model in quite the same way again.

Although it would be inappropriate to elaborate the necessary group theory here, I'll mention that there is a beautiful extension of $SU(5)$ to the slightly larger group $SO(10)$, which permits one to unite all the fermions of a family into a single multiplet [18]. In fact the relevant representation for the fermions is a 16-dimensional spinor representation. Some of its features are depicted in Figure 5. The 16th member of a family in $SO(10)$, beyond the 15 familiar degrees of freedom with a Standard Model family, has the quantum numbers of the right-handed neutrino $N_R$ as mentioned above. This emphasizes again how easy and natural is the extension of the Standard Model to include neutrino masses using the see-saw mechanism.

SO(10): 5 bit register

$$(\pm\,\pm\,\pm\,\pm\,\pm) \;\; : \;\; \underline{even} \; \# \; of \; -$$

$$
\begin{array}{lll}
& (+\,+\,-\,|\,+\,-) & 6 & (\mathrm{u_L,d_L}) \\
10: & (+\,-\,-\,|\,+\,+) & 3 & \mathrm{u_L^c} \\
& (+\,+\,+\,|\,-\,-) & 1 & \mathrm{e_L^c} \\[6pt]
\bar{5}: & (+\,-\,-\,|\,-\,-) & \bar{3} & \mathrm{d_L^c} \\
& (-\,-\,-\,|\,+\,-) & \bar{2} & (\mathrm{e_L},\nu_L) \\[8pt]
1: & (+\,+\,+\,|\,+\,+) & 1 & \mathrm{N_R}
\end{array}
$$

Figure 5: Organization of the fermions in one family, together with a right-handed neutrino degree of freedom, into a single multiplet under $SO(10)$. The components of the irreducible spinor representation, which is used here, can be specified in a very attractive way by using the charges under the $SO(2) \otimes SO(2) \otimes SO(2) \otimes SO(2) \otimes SO(2)$ subgroup as labels. They then appear as arrays of $\pm$ signs, resembling binary registers. There is the rule that one must have an even number of - signs. Strong $SU(3)$ acts on the first three components, weak $SU(2)$ on the final two. The $SU(5)$ quantum numbers are displayed in the left-hand column, the number of entries with each sign-pattern just to the right, and finally the usual Standard Model designations on the far right.

# 4   Unification: Dynamics, and a Big Hint of Supersymmetry [19]

## 4.1   The Central Result

We have seen that simple unification schemes are successful at the level of *classification*; but new questions arise when we consider the dynamics which underlies them.

Part of the power of gauge symmetry is that it fully dictates the interactions of the gauge bosons, once an overall coupling constant is specified. Thus if SU(5) or some higher symmetry were exact, then the fundamental strengths of the different color-changing interactions would have to be equal, as would the (properly normalized) hypercharge coupling strength. In reality the coupling strengths of the gauge bosons in SU(3)×SU(2)×U(1) are observed not to be equal, but rather to follow the pattern $g_3 \gg g_2 > g_1$.

Fortunately, experience with QCD emphasizes that couplings "run". The physical mechanism of this effect is that in quantum field theory the vacuum must be regarded as a polarizable medium, since virtual particle-anti-particle pairs can screen charge. Thus one might expect that effective charges measured at shorter distances, or equivalently at larger energy-momentum or mass scales, could be different from what they appear at longer distances. If one had only screening then the effective couplings would grow at shorter distances, as one penetrates deeper inside the screening cloud. However it is a famous fact[3] that due to paramagnetic spin-spin attraction of like charge vector gluons[20], these particles tend to *antiscreen* color charge, thus giving rise to the opposite effect – asymptotic freedom – that the effective coupling tends to shrink at short distances. This effect is the basis of all perturbative QCD phenomenology, which is a vast and vastly successful enterprise, as we saw in Figure 3.

For our present purpose of understanding the disparity of the observed couplings, it is just what the doctor ordered. As was first pointed out by Georgi, Quinn, and Weinberg[21], if a gauge symmetry such as SU(5) is spontaneously broken at some very short distance then we should not expect that the effective couplings probed at much larger distances, such as are actually measured at practical accelerators, will be equal. Rather they will all have been affected to a greater or lesser extent by vacuum screening and anti-screening, starting from a common value at the unification scale but then diverging from one another at accessible accelerator scales. The pattern $g_3 \gg g_2 > g_1$ is just what one should expect, since the antiscreening or asymptotic freedom effect is more pronounced for larger gauge groups, which have more types of virtual gluons.

The marvelous thing is that the running of the couplings gives us a truly quantitative handle on the ideas of unification, for the following reason. To fix the relevant aspects of unification, one basically needs only to fix two parameters: the scale at which the couplings unite, which is essentially the scale at which the unified symmetry breaks; and their value when they unite. Given these, one calculates three outputs: the three *a priori* independent couplings for the gauge groups SU(3)×SU(2)×U(1) of the Standard Model. Thus the framework is eminently falsifiable. The miraculous thing is, how close it comes to working (Figure 6).

The unification of couplings occurs at a very large mass scale, $M_{\text{un.}} \sim 10^{15}$ Gev. In the simplest version, this is the magnitude of the scalar field vacuum expectation value that spontaneously breaks SU(5) down to the standard model symmetry SU(3)×SU(2)×U(1), and is analogous to the scale $v \approx 250$ Gev for electroweak symmetry breaking. The largeness of this large scale mass scale is important in several ways:

• It explains why the exchange of gauge bosons that are in SU(5) but not in SU(3)×SU(2)×U(1), which re-shuffles strong into weak colors and generically violates the conservation of baryon number, does not lead to a catastrophically quick decay of nucleons. The rate of decay goes as the inverse fourth power of the mass of the exchanged gauge particle, so the baryon-number violating processes are predicted to be far slower than ordinary weak processes, as they had better be.

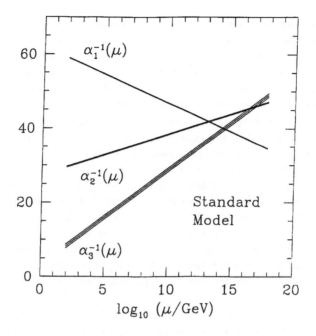

Figure 6: Evolution of Standard Model effective (inverse) couplings toward small space-time distances, or large energy-momentum scales. Notice that the physical behavior assumed for this Figure is the direct continuation of Figure 3, and has the same conceptual basis. The error bars on the experimental values at low energies are reflected in the thickness of the lines. Note the logarithmic scale. The qualitative aspect of these results is extremely encouraging for unification and for extrapolation of the principles of quantum field theory, but there is a definite small discrepancy with recent precision experiments.

• $M_{\rm un.}$ is significantly smaller than the Planck scale $M_{\rm Planck} \sim 10^{19}$ Gev at which exchange of gravitons competes quantitatively with the other interactions, but not ridiculously so. This indicates that while the unification of couplings calculation itself is probably safe from gravitational corrections, the unavoidable logical next step in unification must be to bring gravity into the mix.

• Finally one must ask how the tiny ratio of symmetry-breaking mass scales $v/M_{\rm un.} \sim 10^{-13}$ required arises dynamically, and whether it is stable. This is the so-called gauge hierarchy problem, which I shall discuss in a more concrete form momentarily.

The success of the GQW calculation in explaining the observed hierarchy $g_3 \gg g_2 > g_1$ of couplings and the approximate stability of the proton is quite striking. In performing it, we assumed that the known and confidently expected particles of the Standard Model exhaust the spectrum up to the unification scale, and that the rules of quantum field theory could be extrapolated without alteration up to this mass scale – thirteen orders of magnitude beyond the domain they were designed to describe. It is a triumph for minimalism, both existential and conceptual.

However, on further examination it is not quite good enough. Accurate modern measurements of the couplings show a small but definite discrepancy between the couplings, as appears in Figure 6. And heroic dedicated experiments to search for proton decay did not find it [22]; they currently exclude the minimal SU(5) prediction $\tau_p \sim 10^{31}$ yrs. by about two orders of magnitude.

Given the scope of the extrapolation involved, perhaps we should not have hoped for more. There are several perfectly plausible bits of physics that could upset the calculation, such as the existence of particles with masses much higher than the electroweak but much smaller than the unification scale. As virtual particles these would affect the running of the couplings, and yet one certainly cannot exclude their existence on direct experimental grounds. If we just add particles in some haphazard way things will only get worse: minimal SU(5) nearly works, so the generic perturbation from it will be deleterious. This is a major difficulty for so-called technicolor models, which postulate many new particles in complicated patterns. Even if some *ad hoc* prescription could be made to work, that would be a disappointing outcome from what appeared to be one of our most precious, elegantly straightforward clues regarding physics well beyond the Standard Model.

Fortunately, there is a theoretical idea which is attractive in many other ways, and seems to point a way out from this impasse. That is the idea of supersymmetry [23]. Supersymmetry is a symmetry that extends the Poincare symmetry of special relativity (there is also a general relativistic version). In a supersymmetric theory one has not only transformations among particle states with different energy-momentum but also between particle states of different *spin*. Thus spin 0 particles can be put in multiplets together with spin $\frac{1}{2}$ particles, or spin $\frac{1}{2}$ with spin 1, and so forth.

Supersymmetry is certainly not a symmetry in nature: for example, there is certainly no bosonic particle with the mass and charge of the electron. More generally if one defines the $R$-parity quantum number

$$R \equiv (-)^{3B+L+2S} ,$$

which should be accurate to the extent that baryon and lepton number are conserved, then one finds that all currently known particles are $R$ even whereas their supersymmetric partners would be $R$ odd. Nevertheless there are many reasons to be interested in supersymmetry, and especially in the hypothesis that supersymmetry is effectively broken at a relatively low scale, say $\approx$ 1 Tev. Anticipating this for the moment, let us consider the consequences for running of the couplings.

The effect of low-energy supersymmetry on the running of the couplings was first considered long ago [24], well before the discrepancy described above was evident experimentally. One might have feared that such a huge expansion of the theory, which essentially doubles the spectrum, would utterly destroy the approximate success of the minimal SU(5) calculation. This is not true, however. To a first approximation, roughly speaking because it is a space-time as opposed to an internal symmetry, supersymmetry does not affect the group-theoretic structure of the unification of couplings calculation. The absolute rate at which the couplings run with momentum is affected, but not the relative rates. The main effect is that the supersymmetric partners of the color gluons, the gluinos, weaken the asymptotic freedom of the strong interaction. Thus they tend to make its effective coupling decrease and approach the others more slowly. Thus their merger requires a longer lever arm, and the scale at which the couplings meet increases by an order of magnitude or so, to about $10^{16}$ Gev. Also the common value of the effective couplings at unification is slightly larger than in conventional unification ($\frac{g_{un}^2}{4\pi} \approx \frac{1}{25}$ $versus$ $\frac{1}{40}$). This increase in unification scale significantly reduces the predicted rate for proton decay through exchange of the dangerous color-changing gauge bosons, so that it no longer conflicts with existing experimental limits.

Upon more careful examination there is another effect of low-energy supersymmetry on the running of the couplings, which although quantitatively small has become of prime interest. There is an important exception to the general rule that adding supersymmetric partners does not immediately (at the one loop level) affect the relative rates at which the couplings run. This rule works for particles that come in complete SU(5) multiplets, such as the quarks and leptons (which, since they don't upset the full SU(5) symmetry, have basically no effect) or for the supersymmetric partners of the gauge bosons, because they just renormalize the existing, dominant effect of the gauge bosons themselves. However there is one peculiar additional contribution, from the supersymmetric partner of the Higgs doublet. It affects only the weak SU(2) and hypercharge U(1) couplings. (On phenomenological grounds the SU(5) color triplet partner of the Higgs doublet must be extremely massive, so its virtual exchange is not important below the unification scale. *Why* that should be so, is another aspect of the hierarchy problem.) Moreover, for slightly technical reasons even in the minimal supersymmetric model it is necessary to have two different Higgs doublets with opposite hypercharges[c]. The main effect of doubling the number of Higgs fields and including their supersymmetric partners is a sixfold enhancement of the asymmetric Higgs field contribution to the running of weak and hypercharge couplings. This causes a small, accurately calculable change in the calculation. From Figure 7 you see that it is a most welcome one. Indeed, in the minimal implementation of supersymmetric unification, it puts the running of couplings calculation right back on the money [25].

Since the running of the couplings with scales depends only logarithmically on the mass scale, the unification of couplings calculation is not terribly sensitive to the precise scale at which supersymmetry is broken, say between 100 Gev and 10 Tev. (To avoid confusion later, note that here by "the scale at which supersymmetry is broken" I mean the typical mass splitting between Standard Model particles and their supersymmetric partners. The phrase is frequently used in a different sense, referring to the largest splitting between supersymmetric partners in the entire world-spectrum; this could be much larger, and indeed in popular models it almost invariably is. The ambiguous terminology is endemic in the literature; fortunately, the meaning is usually clear from the context.) There have been attempts to push the calculation further, in order to address this question of the supersymmetry breaking scale, but they are controversial. For example, comparable uncertainties arise from the splittings among the very large number of particles with masses of order the unification scale, whose theory is poorly developed and unreliable. Superstring theory suggests [26] many possible ways in which the simple calculation described here might go wrong[d]; if we take the favorable result of this calculation at face value, we must conclude that none of them happen.

In any case, if we are not too greedy the main points still shine through:

• If supersymmetry is to fulfill its destiny of elucidating the hierarchy problem in a straightforward way, then the supersymmetric partners of the known particles cannot be much heavier than the $SU(2) \times U(1)$ electroweak breaking scale, i.e. they should not be beyond the expected reach of LHC.

• If we assume this to be the case then the meeting of the couplings takes place in the simplest minimal models of unification, to adequate accuracy, without further assumption. This is a most remarkable and non-trivial fact.

---

[c]Perhaps the simplest, though not the most profound, way to appreciate the reason for this has to do with anomaly cancelation. The minimal spin-1/2 supersymmetric partner of the Higgs doublet is chiral and has non-vanishing hypercharge, introducing an anomaly. By including a partner for the anti-doublet, one cancels this anomaly.

[d]Indeed, in the simplest superstring-inspired models it is not entirely easy to accommodate the 'low' value of the unification scale compared to the Planck scale.

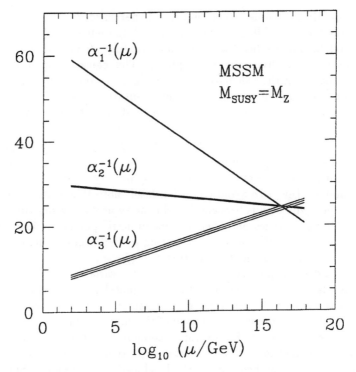

Figure 7: Evolution of the effective (inverse) couplings in the minimal extension of the Standard Model, to include supersymmetry. The concepts and qualitative behaviors are only slightly modified from Figure 6 (a highly non-trivial fact!) but the quantitative result is changed, and comes into adequate agreement with experiment. I would like to emphasize that results along these lines were published well before the difference between Figure 6 and Figure 7 could be resolved experimentally, and in that sense one has already derived a successful *prediction* from supersymmetry.

## 4.2 Implications

The preceding result, taken at face value, has extremely profound implications:

• Quantum field theory, and specifically its characteristic vacuum polarization effects leading to asymptotic freedom and running of the couplings, continue to work quantitatively up to energy scales many orders of magnitude beyond where they were discovered and established.

I would like to emphasize also some negative implications of this: there are things that might have, but do *not*, happen. It might have happened that the known particles are some complicated composites of more elementary objects, or that many additional strong couplings appeared at higher energies (technicolor), or that additional dimensions became dynamically active, or that particle physics simply dissolved into some amorphous mess. Unless Figure 7 is a cruel joke on the part of mother Nature, none of this happens, or at least the complications are in a strong, precise sense walled off from the Standard Model and the dynamical evolution of its couplings.

• Supersymmetry, in its virtual form, has already been discovered.

## 4.3 Why Supersymmetry is a Good Thing

Thus has Nature spoken, in a promissory whisper. Many of us are seduced, because She is telling us something we want to hear:

• You will notice that we have made progress in uniting the gauge bosons with each other, and the various quarks and leptons with each other, but not the gauge bosons with the quarks and leptons. It takes supersymmetry – perhaps spontaneously broken – to make this feasible.

• Supersymmetry was invented in the context of string theory, and seems to be necessary for constructing consistent string theories containing gravity (critical string theories) that are at all realistic.

• Most important for present purposes, supersymmetry can help us to understand the vast disparity between weak and unified symmetry breaking scales mentioned above. This disparity is known as the gauge hierarchy problem. It actually raises several distinct problems, including the following. In calculating radiative corrections to the $(\text{mass})^2$ of the Higgs particle from diagrams of the type shown in Figure 8 one finds an infinite, and also large, contribution. By this I mean that the divergence is quadratic in the ultraviolet cutoff. No ordinary symmetry will make its coefficient vanish. If we imagine that the unification scale provides the cutoff, we find that the radiative correction to the $(\text{mass})^2$ is much larger than the final value we want. (If the Higgs field were composite, with a soft form factor, this problem might be ameliorated. Following that road leads to technicolor, which as mentioned before seems to lead us far away from our best source of inspiration.) As a formal matter, one can simply cancel the radiative correction against a large bare contribution of the opposite sign, but in the absence of some deeper motivating principle this seems to be a horribly ugly procedure. Now in a supersymmetric theory for any set of virtual particles circulating in the loop there will also be another graph with their supersymmetric partners circulating. If the partners were accurately degenerate, the contributions would cancel. Otherwise, the threatened quadratic divergence will be cut off only at virtual momenta such that the difference in $(\text{mass})^2$ between the virtual particle and its supersymmetric

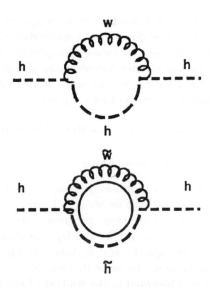

Figure 8: Contributions to the Higgs field self-energy. These graphs give contributions to the Higgs field self-energy which separately are formally quadratically divergent, but when both are included the divergence is removed. In models with broken supersymmetry a finite residual piece remains. If one is to obtain an adequately small finite contribution to the self-energy, the mass difference between Standard Model particles and their superpartners cannot be too great. This – and essentially only this – motivates the inclusion of virtual superpartner contributions in Figure 7 beginning at relatively low scales.

partner is relatively negligible. Thus we will be assured adequate cancelation if and only if supersymmetric partners are not too far split in mass – in the present context, if the splitting is not much greater than the weak scale. This is (a crude version of) the most important *quantitative* argument which suggests the relevance of "low-energy" supersymmetry.

• Supersymmetric field theories have many special features, which make them especially interesting, and perhaps promising, phenomenologically.

I cannot be very specific about this here, both because there are as yet no canonical models and because the subject is excessively technical, but let me just mention some appropriate concepts: radiative $SU(2) \times U(1)$ breaking associated with the heavy top quark; doublet-triplet splitting mechanisms; approximate flat directions for generating large mass hierarchies. Supersymmetric models also have additional mechanisms for neutral flavor-changing processes and CP violation, which are dangerously large generically, but in appropriate models can be suppressed down to a level which is interesting – but not *too* interesting – experimentally.

All this provides, in my opinion, a very good specific brief for optimism about the future of experimental particle physics exploring the high-energy frontier, and also – with somewhat less certainty – the frontier of small exotic flavor-changing and CP violating processes. We can already discern, at the limit of our vision, the shores of a strange new world not too far away, where we can realistically hope to land and explore.

## 5   Some Additional Remarks

I have completed the main case I wanted to make, the case for low-energy supersymmetry as an outstanding development in particle physics that will soon come to spectacular fruition. I hope I did justice to its logical force. Though the case is drawn entirely from what we would call, in the American legal system, "circumstantial evidence", I believe it is very strong indeed.

Now I would like to close with a few brief comments about other important challenges for the future of particle physics.

One big challenge is to better understand string theory. I will not say much about string theory here, partly because my colleague David Gross will be discussing it at length, but also partly to emphasize how much one can say without invoking it. The principles involved in the principal line of argument I described above not only could have been, but historically in fact were, substantially in place before the string renaissance of 1984. Indeed, it remains a great challenge to develop string theory to the point where it will add useful, concrete insight into the questions raised in part 2. At present, I think it is fair to say that among these questions, only the ultraviolet problem of quantum gravity has been reasonably convincingly advanced – and while of course that is a profound problem, as I discussed it is rather an academic one. It may be more realistic and appropriate, for the forseeable future, to try to address much more qualitative questions. Can the theory tell us about the pattern of supersymmetry breaking: how is the spectrum of gauginos, squarks, and sleptons ordered? How are various rare and exotic flavor-violating processes, which arise generically in supersymmetric models, avoided? Can the theory tell us why space-time is effectively 3+1 dimensional? Can it tell us why the cosmological term is so nearly zero? Finally: can the theory be adequately (and, one hopes, beautifully) defined, apart from various limits and perturbative expansions? The gap between promise and delivery remains very wide.

A second big challenge is to relate ideas in modern particle physics to phenomena in the natural world, away from accelerators. One set of applications involves the noble enterprise of learning to calculate more and more with QCD, relating the fundamental microscopic theory to nuclear physics and to possible exotic phases of matter including pion or (especially) kaon condensation, strange matter, or quark-gluon plasma, produced in neutron stars or the big bang at $T \sim 10^{12}$K. The other, more properly 'modern', set of applications, those using ideas beyond the Standard Model, must, unless we are very lucky (naked singularities?), involve still earlier moments of the big bang. Since spontaneous symmetry breaking plays such a central role both in the Standard Model and its most attractive extensions, and we expect most symmetries to be restored at sufficiently high temperatures, the possibility of cosmic phase transitions is quite compelling. These can potentially leave various signatures including relics such as cosmic strings, gravity waves, or several others. Most important, such phase transitions might trigger an inflationary epoch. Existing models of inflation for the most part make at best qualitative contact with genuine world-models of particle physics. We desperately need a convincing implementation. Or perhaps a replacement? Let me remind you that the triumphs of inflation are all qualitative, so that the possibility of an alternative explanation of the phenomena does not seem altogether absurd; and that existing models of inflation make heavy use of the cosmological term, of which our understanding is quite problematic.

Also in connection with cosmology I would like to mention a circle of ideas which though it can be made to sound quite fantastic I think is actually deeply implicit in much current thinking about particle physics and cosmology.

The axion is perhaps the most well-motivated and studied exemplar of a family of related fields including familons, dilatons, and moduli fields that in one way or another embody the idea that what we ordinarily consider 'constants' might not be fundamental parameters fixed in the very formulation of the laws of Nature, but rather can be considered usefully as dynamical entities. In a theory with only one fundamental, dimensional, parameter, such as superstring theory appears to be, there is clearly a sense in which all dimensionless couplings are dynamical variables. They might nevertheless behave effectively as constants either if it costs a very large energy-density to excite them at all (*e.g.* massive fields) or if ordinary matter couples very weakly to long-wavelength fluctuations of the field (*e.g.* stiff light fields with derivative couplings).

In the latter case one might anticipate long-range forces mediated by the exchange of the field. I believe that experiments to look for such forces are among the most fundamental which can presently be attempted. They address in a concrete way the question: are the constants of Nature uniquely determined to be what we observe by dynamical laws, or are they 'frozen accidents' imprinted at the Big Bang? Or are they presently relaxing towards some more favorable value$^e$? Note that although a rapidly (on cosmological time scales) oscillating field represents non-relativistic matter, a sufficiently slowly varying field might appear as a contribution to the effective cosmological term.

If these ideas are along the right lines, it could be misguided to seek a unique Lorentz-invariant 'vacuum' state as a model for our present world. One might instead be required, at a fundamental level, to seek the boundary conditions for particle physics from cosmology (and *vice versa*).

To address this sort of question, one will probably need to relate string theory to dynamical space-time models in ways that are beyond current understanding, which in particular do not respect supersymmetry. For an attempt to get started along these lines, see [27].

A third big challenge, or better opportunity, is to connect the *ideas*, as opposed to the specific world-models, used in particle physics to those used in other parts of physics, especially many-body theory. This enterprise has a long and glorious history, featuring flow in both directions [28]. At present, the most obviously attractive area for such connections is in description of the complex of phases of matter associated with the quantum Hall effect. Ideas of (abelian and non-abelian!) gauge theory, confinement, use of holomorphic functions and non-commuting coordinates, conformal field theory, instantons, skyrmions, and others find quite non-trivial use in describing these phases. New forms of quantum symmetry are exhibited in some of these phases; in particular what I call obstructed symmetry. This is symmetry which is valid for the ground state, but which can be obstructed by long-range fields occurring in certain sectors of the theory, *i.e.* symmetry whose extent which varies with the state, so that a given state will generally not be covariant with respect to the full symmetry of the ground state, but only to a variable subgroup [29-30]. I find it quite plausible that other logically natural but unfamiliar phenomena, specifically including the quasi-nonlocality of Cheshire charge [31], will in due course make their appearance in real materials, either in the quantum Hall complex or elsewhere.

To summarize, both the internal logic of particle physics as outlined in previous sections, and this brief inventory of challenges and opportunities, show that the theory of high-energy physics is very far from being complete and satisfactory. I think there should be a moratorium on loose talk about 'The Theory of Everything' and its evil twin 'The End of Physics' while the theory of just about every particular thing, once we venture beyond the Standard

---

$^e$The standard picture of axion production in the Big Bang is essentially an example of this: the axion field drops out of equilibrium as the Universe cools from $10^{12}$ to about 1 Gev, then relaxes towards the dynamically favored, approximately P and T conserving, value.

Model, is grossly inadequate. As my mother might say, "Such problems you should have."

## Acknowledgments

I thank Keith Dienes for supplying Figures 6 and 7, and Chris Kolda and John March-Russell for valuable assistance in the preparation of this manuscript. Research supported in part by DOE grant DE-FG02-90ER40542.

## References

1. For recent reviews of the Standard Model, see References 4 and 5.
2. After several partial and tentative proposals, the $SU(2) \times U(1)$ electroweak theory took on in its essentially modern form in: S. Weinberg, *Phys. Rev. Lett.* **19**, 1264 (1967); A. Salam, in *Elementary Particle Physics*, ed. N. Svartholm (Almqvist and Wiksells, Stockholm, 1968), p. 367; S. Glashow, J. Iliopoulos, and L. Maiani, *Phys. Rev.* D **2**, 1285 (1970).
3. After several partial and tentative proposals, the $SU(3)$ strong interaction theory took on its essentially modern form in: D. Gross and F. Wilczek, *Phys. Rev.* D **8**, 3633 (1973); S. Weinberg, *Phys. Rev. Lett.* **31**, 494 (1973); H. Fritzsch, M. Gell-Mann, and H. Leutwyler, *Phys. Lett.* B **47**, 365 (1973). Not coincidentally, the key discovery that allowed one to connect the abstract gauge theory to experiments, asymptotic freedom, was first demonstrated just prior to these papers, in: D. Gross and F. Wilczek, *Phys. Rev. Lett.* **30**, 1343 (1973); H. D. Politzer, *Phys. Rev. Lett.* **30**, 1346 (1973).
4. LEP Electroweak Working Group, preprint CERN-PPE/96-183 (Dec. 1996).
5. M. Schmelling, preprint MPI-H-V39, hep-ex/9701002. Talk given at the 28th International Conference on High-energy Physics (ICHEP 96), Warsaw, Poland, 25-31 July 1996.
6. G. 't Hooft, *Phys. Rev. Lett.* **37**, 8 (1976); C. Callan, R. Dashen, and D. Gross, *Phys. Lett.* B **63**, 334 (1976); R. Jackiw, C. Rebbi, *Phys. Rev. Lett.* **37**, 172 (1976).
7. R. Peccei and H. Quinn, *Phys. Rev. Lett.* **38**, 1440 (1977), *Phys. Rev.* D **16**, 1791 (1977).
8. S. Weinberg, *Phys. Rev. Lett.* **40**, 223 (1978); F. Wilczek, *Phys. Rev. Lett.* **40**, 279 (1978).

9. J. Preskill, M. Wise, and F. Wilczek, *Phys. Lett.* B **120**, 127 (1983); L. Abbott and P. Sikivie, *Phys. Lett.* B **120**, 133 (1983); M. Dine and W. Fischler, *Phys. Lett.* B **120**, 137 (1983).

10. The existence of the infrared fixed point was first discussed in: C. Hill, *Phys. Rev.* D **24**, 691 (1981). More recent examinations including supersymmetry appear in: V. Barger, M. Berger and P. Ohmann, *Phys. Rev.* D **47**, 1093 (1993); P. Langacker and N. Polonsky, *Phys. Rev.* D **47**, 4028 (1993); M. Carena, S. Pokorski and C. Wagner, *Nucl. Phys.* B **406**, 59 (1993).

11. C. Jarlskog, *Phys. Rev. Lett.* **55**, 1039 (1985).

12. J. Bahcall, *et al*, Nature **375**, 29 (1995).

13. G. Fogli, E. Lisi, and D. Montanino, *Phys. Rev.* D **49**, 3626 (1994).

14. C. Athanassopoulos, *et al*, *Phys. Rev. Lett.* **75**, 2650 (1995).

15. M. Gell-Mann, P. Ramond, and R. Slansky, in *Supergravity*, ed. P. van Neiuwenhuizen and D. Freedman (North Holland, Amsterdam, 1979), p. 315; T. Yanagida, Proc. of the Workshop on Unified Theory and Baryon Number in the Universe, eds. O. Sawada and A. Sugamoto (KEK, 1979).

16. S. Hawking, *Phys. Rev.* D **14**, 2460 (1976).

17. H. Georgi and S. Glashow, *Phys. Rev. Lett.* **32**, 438 (1974).

18. H. Georgi, in *Particles and Fields - 1974*, ed. C. Carlson (AIP press, New York, 1975).

19. See S. Dimopoulos, S. Raby and F. Wilczek, Phys. Today **44**, 25 (1991) and references contained therein.

20. N. Nielsen, Am. J. Phys. **49**, 1171 (1981); R. Hughes, *Nucl. Phys.* B **186**, 376 (1981).

21. H. Georgi, H. Quinn, and S. Weinberg, *Phys. Rev. Lett.* **33**, 451 (1974).

22. See for example G. Blewitt, *et al*, *Phys. Rev. Lett.* **55**, 2114 (1985), and the latest Particle Data Group compilations.

23. A very useful introduction and collection of basic papers on supersymmetry is S. Ferrara, *Supersymmetry* (2 vols.) (World Scientific, Singapore 1986). Another excellent standard reference is N.-P. Nilles, Phys. Reports **110**, 1 (1984). See also [26].

24. S. Dimopoulos, S. Raby, and F. Wilczek, *Phys. Rev.* D **24**, 1681 (1981).

25. J. Ellis, S. Kelley, and D. Nanopoulos, *Phys. Lett.* B **260**, 131 (1991); U. Amaldi, W. de Boer, and H. Furstenau, *Phys. Lett.* B **260**, 447 (1991); for more recent analysis see P. Langacker and N. Polonsky, *Phys. Rev.* D **49**, 1454 (1994).

26. K. Dienes, IASSNS-HEP-95/97, hep-th/9602045 (Feb 1996), *Phys. Reports* (in press).

27. F. Larsen and F. Wilczek, IASSNS-HEP-96/108 (Oct. 1996), *Phys. Rev. D* (in press).

28. I plan to write about this in a series of Reference Frame articles for Physics Today.

29. A. Schwarz, *Nucl. Phys.* B **208**, 141 (1982).

30. P. Nelson and S. Coleman *Nucl. Phys.* B **237**, 1 (1984).

31. M. Alford, *et al*, *Phys. Rev. Lett.* **64**, 1632 (1990) [Erratum: 65 (1990) 668]; *Nucl. Phys.* B **349**, 414 (1991).

# The Future of Particle Physics

DAVID J. GROSS[a]
*Joseph Henry Laboratories*
*Princeton University*
*Princeton, New Jersey 08544*

## 1 Introduction

It is a pleasure and an honor to speak at this wonderful symposium devoted to the future of physics. The subject of my talk is quite apporpriate, for I shall discuss the future of elementary particle phsyics.

Does elementary particle physics retain the scientific vitality that has characterized it for the last fifty years? At the very moment that elementary particle physics, after fifty years of intense experimental and theoretical effort, has succeeded in developing a comprehensive theory of the known forces—the standard model—claims are made that the field is dead. The very success of the standard model has led some to experience boredom. The experimental facilities needed to explore beyond what we now know are bigger, more expensive and more difficult to exploit. The time scale of experimental particle physics has perhaps doubled, so that it takes approximately fifteen years from the planning of an experiment to its completion. The size of the collaborations involved in these experiments has also greatly expanded with some unfortunate consequences. On a more fundamental level, we can now identify new scales of energy, where new physics surely occurs, but which at the moment seem frustratingly unattainable.

In view of all this, the biggest danger to particle physics is that it will dry up and that will cease to attract the best and the brightest young people. A scientific field requires at least two things to remain vital. First, it requires good questions, interesting questions, important questions, and accessible questions. Second, it requires new experimental instruments and techniques that can be used to probe and answer these questions.

Below I review the state of particle physics and argue that the ingredients for the continued vitality of the field exist, that the questions that we can ask at this point in time are as exciting as they have ever been, if not more so. Furthermore in the coming decade the instruments necessary to address some of these questions will be built, and exciting discoveries are likely to be

---

[a]Present Address: Institute For Theoretical Physics, University of California, Santa Barbara, California

98

made. Finally we have the beginning of an exciting new theory which portends
to alter the conceptual structure of microscopic physics, to revolutionize our
notions of space and time, and to provide the framework for a unified theory
of all the forces of nature.

I start by reviewing, very schematically, the deep conceptual lessons that
we have learned over the past fifty years. Then I enumerate some of the
questions that we can now ask and hope to answer and address the challenges
that confront us—immediate and long term. Finally, I discuss the status of
string theory.

## 2 What Have We Learned?

Elementary particle physics began in earnest after World War II with the
advent of modern accelerators that could probe the microscopic structure of
matter to incredibly small distances. By the middle of the 1970's the many
discoveries made by these instruments had prompted theorists to develop a
comprehensive theory of the fundamental constituents of matter and the laws
that govern their interactions— the "standard model." This theory describes
the forces of electromagnetism, the weak interaction responsible for radioactivity, and the strong nuclear force that governs the structure of nuclei, as
consequences of local (gauge) symmetries. These forces act on the fundamental constituents of matter, which have been identified as pointlike quarks and
leptons. In the following decades this theory has been subjected to precise
tests, and its theoretical structure has been greatly developed and understood.

What have we learned in fifty years? There are four important lessons,
that I feel are more fundamental than the precise details of the standard model:

- Quantum field theory works;

- The secret of nature is symmetry;

- The standard model teaches us little about the fundamental theory of
  physics;

- We have learned many new questions.

### 2.1 Quantum Field Theory Works

The first thing we have learned is that quantum field theory—based on the
principles of quantum mechanics, relativity and local fields—provides the theoretical framework for understanding fundamental physics. It is the basis of
the standard model. As such we have tested its basic tenets down to distances

of the order of $10^{-17}$cm. with increasing precision. The success of quantum field theory is quite remarkable, especially if we remember that over the initial period of the development of quantum field theory (in the 1930's) it was under constant attack. Most of the founders of quantum field theory, from Heisenberg and Dirac to their successors, believed that this framework was inadequate to describe microscopic physics. They were skeptical that the classical notion of a field, introduced by Faraday and Maxwell to describe electromagnetism, was robust enough to apply to the quantum domain of nuclear interactions. They felt that the many technical and conceptual difficulties that arose in quantizing classical field theory indicated that a fundamental revision was required at small distances.

Today we see no reason why quantum field theory will not continue to work until we get down to unbelievably short distances of order the Planck length of $10^{-33}$cm., where quantum gravity becomes a strong force and the fluctuations of spacetime are large. In this domain we suspect that quantum field theory is no longer adequate. That is not to say that there before the Planck length substantial changes in the standard model or new insights into the structure of the quantum field theory will not be required; but rather that quantum field theory as the basic theoretical framework for understanding nature probably works until this scale of distances. At the Planck length quantum field theory will probably fail since it does not seem to be able to describe quantum gravity. It will most likely be replaced by something like string theory. Thus, contrary to the belief of it's originators, quantum field theory, embodied in the standard model, incorporates all of known physics in a reductionist sense, from the microscopic regime to the edge of the universe—60 orders of magnitude. We have never had a more successful theory of nature.

## 2.2   The Secret of Nature is Symmetry

The main lesson of twentieth century physics is that *nature's laws are based on symmetry.* Einstein's great advance in 1905 was to put symmetry first, to regard the symmetry principle as the primary feature of nature that constrains the allowable dynamical laws. Thus the transformation properties of the electromagnetic field were not to be derived from Maxwell's equations, as Lorentz did, but rather were consequences of the symmetry of spacetime, which largely dictate the form of Maxwell's equations. Ten years later this point of view scored a spectacular success with Einstein's construction of general relativity. The principle of equivalence, a principle of local symmetry—the invariance of the laws of nature under local changes of the spacetime coordinates—dictated the dynamics of gravity, of spacetime itself. Much of the subsequent progress that we have made has been based on discovering and understanding new symmetries of nature. Indeed all of our theories of fundamental physics, including the standard model's laws of interaction as well as Einstein's theory of gravity, are based on local symmetries.

Our understanding of the meaning of symmetry has matured and grown in ways that were unimaginable when the notion of symmetry was still fresh, as it was at the outset of this century. Originally the focus was on global geometric symmetries of spacetime such as invariance of the laws of physics under spacetime translations or rotations. These traditional symmetries are regularities of the laws of motion and are formulated in terms of physical events; the application of the symmetry transformation yields a different, yet equivalent, physical situation. Today the focus has shifted to local, *gauge symmetries*, which act locally on an internal space. Gauge symmetries are formulated only in terms of the laws of nature; the application of the symmetry transformation merely changes our description of the same physical situation. Gauge invariance is a symmetry of our description of nature, yet it underlies dynamics. As Yang has stated: *Symmetry dictates interaction.* The requirement of gauge symmetry leads to specific dynamics and requires the existence of specific particles the mediate the forces. In the standard model the carriers of these forces are the gluons (the glue that holds the nucleus together), the photon (or light rays) as well as the $W$ and $Z$ particles observed a few years ago.

The symmetries that we observe in nature are seen indirectly. They are not obvious or manifest. That is why it takes so long to find them. Indeed the second lesson that we have learned is that *Nature's laws are based on symmetry, but the texture of the world that we see around us is a consequence of mechanisms of symmetry breaking or of confinement that hide the underlying symmetry from us.* These effects give rise to much more interesting and complex physical states in which the symmetry is not manifested. The spontaneous symmetry breaking of global and local gauge symmetries is a recurrent theme in modern theoretical physics. In quantum mechanical systems with an infinite number of degrees of freedom (quantum field theory where each point in space is associated with an independent dynamical entity; or large macroscopic bodies, which can be regarded as containing an infinite number of atoms), it is often the case that the stable ground state of the system does not exhibit the symmetry of the laws of nature. Magnetism is the prime example of this, wherein the atomic spins are aligned in a certain direction, producing macroscopic magnetism, although the laws of nature are invariant under spatial rotations. Such spontaneous symmetry breaking is responsible for magnetism, superconductivity, the structure of the unified electro-weak theory and more. Indeed the search for new symmetries of nature is based on the possibility of symmetry breaking, for a new symmetry that we discover must be somehow broken otherwise it would have been apparent long ago.

Much of the development of the standard model and the elucidation of its dynamics was based on discovering these symmetries and discovering the mechanisms of symmetry breaking and confinement. The laws of interaction of the standard model, the laws of force, are based on local gauge symmetry. But the other component of the standard model is the content of the matter, the quarks and the leptons that come in repeated families with a totally bizarre mass spectrum. The fact that this matter is fermionic (i.e., obeys Fermi-Dirac statistics) and that it possesses a left-handedness is another key ingredient and important component of the standard model. But the origin of the breaking of parity, as well as time reversal and charge conjugation, remains a mystery, together with the couplings of the fermions that determine their masses. It is in this sector that the standard model is unable to explain much and that most of the arbitrary parameters appear.

Current theoretical exploration in the search for further unification of the forces of nature, including gravity, is largely based on the search for new symmetries of nature. Theorists speculate on larger and larger local symmetries and more intricate patterns of symmetry breaking in order to unify the separate interactions.

## 2.3   Effective Dynamics

The third lesson that we have learned from the success of the standard model and from our understanding of quantum field theory is that the standard model is not a fundamental theory at all. In fact it teaches us little as to what the fundamental theory of physics is.

This lesson is based on the notion of *effective dynamics*. The decoupling of physical phenomena at different scales of energy is an essential characteristic of nature. QCD is the theory of quarks whose strong interactions are mediated by gluons. This is the appropriate description at energies of billions of electron volts. However, if we wish to describe the properties of ordinary nuclei, at energies of millions of electron volts, we employ instead an *effective theory* of nucleons, composites of the quarks, whose interactions are mediated by other quark composites—mesons. Similarly, in order to discuss the properties of ordinary matter made of atoms at energies of a few electron volts we can treat the nuclei as pointlike particles, ignore their internal structure, and take into account only the electromagnetic interactions of the charged nuclei and electrons. These low energy effective theories can in principle, and sometimes in practice, be derived from the high energy theory; but most of the low energy dynamics is independent of the details of the latter. It is this feature of nature that makes it possible to understand a limited range of physical phenomena without having to understand everything at once. The same philosophy can be applied to the standard model itself.

Imagine that we have a unified theory (for example, string theory) whose characteristic energy scale, $\Lambda$, is very large or whose characteristic distance scale, $\hbar c/\Lambda$, is very small (say the Planck length of $10^{-33}$cm.). Assume further that just below this scale the theory can be expressed in terms of local field variables. As to what happens at the unification scale itself we assume nothing, except that just below this scale the theory can be described by a local quantum field theory. String theory does provide us with an example of such a unified theory, which includes gravity and can be expressed by local field theory at distances much larger than the Planck length. Even in the absence of knowledge regarding the unified theory we can write the most general quantum field theory. It has an infinite number of parameters describing all possible fields and all possible interactions. We also assume that all the dimensionless couplings that characterize the theory at energy $\Lambda$ are of order one (what else could they be?). Such a theory is useless to describe the physics at high energy, since it involves an infinite number of parameters. However, at low energies, of order $E$, the effective dynamics, the effective Lagrangian, that describes physics up to corrections of order $E/\Lambda$ will be parameterized by a finite number of couplings. This is sometimes expressed in terms of how the various couplings run with energy. We start at $\Lambda$ with whatever the final unified theory is, and at low energies all but a finite number of the possible interactions die away, never to be observed (except for gravity which happens to have the advantage of coupling to mass so that we can construct large objects like planets and scatter them off each other). If we demand further that the theory at the scale $\Lambda$ contain the local gauge symmetry that we observe in nature, then the effective low energy theory will be described by the standard model up to terms that are negligible by inverse powers of the large scale compared to the energy that we observe. These extra interactions will give rise to weak effects, such as gravity or proton decay. But these are very small and unobservable at low energy.

Thus the standard model is the inevitable consequence of any unified theory, any form of the final theory, as long as it is local at the very high energy scale and contains the observed low energy symmetries. It is pleasing that we understand why the standard model emerges at low energy. But from the point of view of the unified theory that surely awaits us at very high energy, it is disappointing since our low energy theory tells us little about what the final theory can be. When I was a graduate student, anyone who would have dreamt that there would exist by 1975 a theory with 19 parameters, that would explain the strong and weak and electromagnetic interactions, would have thought that that would be the final theory. Now we know that it is far from the final theory. In fact it tells us very little about the final theory or the next theory.

## 2.4  Questions

The fourth and perhaps the most important lesson that we have learned are new questions. New questions are essential for the continued vitality of science. The most important product of our knowledge is ignorance—educated ignorance. One must be educated in order to be able to ask the right questions.

In physics there are three stages of exploration and understanding. At first, as one discovers a new phenomenon in nature one asks the question "What?", what is the phenomenon? This stage of exploration is primarily experimental. Once this stage is passed, physicists ask the question, "How?", how does it work? This stage is a joint effort of experimenters and theorists trying to discover and understand the regularities of the physical phenomenon and to codify them in a mathematical formalism. But after one has understood what the phenomenon is and how it works, then one begins to ask, "Why?", why is it so? This stage is primarily a theoretical effort. It often leads to new insights that suggest new experiments that discover new phenomena, and thus the cycle begins anew. In the case of elementary particle physics, we are now at the third stage. We have constructed a successful theory of all the forces of nature that does explain the observed regularities. Now that we understand how it works, we are beginning to ask why questions. The standard model certainly deals with the what and how questions, but it cannot answer most of the why questions:

- Why the specific symmetry group that underlies the standard model? This group and the observed matter multiplets cry out to be unified in a larger, simpler group. How will that take place?

- How and why does the symmetry breaking actually occur and what drives it? This is partly a how question, whose answer we think we know; yet many of the details are not clear and have not yet been explored experimentally.

- Why are there three families of quarks and leptons? Why is there fermionic matter at all? What explains the strange couplings and the bizarre mass spectrum of the quarks and leptons?

- Why and how is CP symmetry ( the simultaneous reversal of all charges and mirror reflection) violated?

- Do neutrinos get a mass, and what is the mass spectrum of the neutrinos?

- Why is there an enormous disparity between the fundamental Planck scale and the electroweak scale of the standard model? This is the modern version of the *large numbers problem* posed by Dirac, who wondered why the proton mass was so much less than the Planck mass.

- Then there are a whole host of questions which have to do with the structure of space and time, which is now a concern of particle physics since the scale of unification seems close to the Planck scale. We do not have a satisfactory theory of quantum gravity, and the ordinary field theoretic approach probably fails. We must therefore ask how quantum gravity works? Even questions previously regarded as metaphysical, such as why are there four dimensions of space and time and why are time and space different, now appear to be addressable.

- Finally, the global structure of spacetime and cosmology (the history of the universe) has become a major concern of particle physics. We wish to understand why spacetime is so flat, why is the cosmological constant (the background energy density of the universe) is so close to zero if not exactly zero, and how did the universe begin? Indeed theoretical progress in particle physics is essential to the understanding of the history of the universe, and cosmological observations provide new experimental clues for fundamental physics.

This is an impressive list of wonderful and interesting questions that we believe that we can start to address both theoretically and experimentally.

## 3 What Are the Challenges?

I now outline the challenges I see for particle physics in the near and far future.

### 3.1 Within the Standard Model

The standard model, although a rather well defined theory that works very well, has many open challenges and problems. Perhaps the most complete part of the standard model is QCD, the theory of the strong nuclear force, a totally well defined theory with essentially no free parameters whose dynamics we understand but which we do not know how to solve. At high energies, due to asymptotic freedom, the theory can be treated by perturbative techniques. One important task is to improve the calculation of strong interaction processes at high energy. This is essential for controlling the background for future experiments. But more important is the task of gaining analytic control over the mechanism of quark confinement into ( hadronic) bound states. Perhaps we can even dream of solving the theory, or at least developing controllable approximations. We have a theory; we should solve it.

There are also new and fascinating aspects of strong interaction physics that are now accessible with new accelerators that have just turned on or which will turn on shortly. The electron-proton collider at DESY in Germany can explore the structure of the proton. The new heavy ion facility, RHIC, that will soon be built at Brookhaven, will make possible the search for signs of new phases of QCD that should exist when we heat or compress hadronic matter. This fascinating physics deserves to be thoroughly explored both theoretically and experimentally.

As for the electroweak theory, which is less of a complete theory, there are many things to explore and explain. The precision tests of the standard model at the electron-positron collider at CERN (LEP), at the Tevatron at Fermilab, and and at SLAC will continue and are extremely important. The $B$ factories that are now being built in Japan and at Stanford will explore $CP$ violating effects for the heavy quarks, where they might be substantial and perhaps lead us to understand the origin of $CP$ violation. With luck we might discover, either using solar neutrino detectors or observing neutrino oscillations, whether neutrinos do have masses and whether they can have something to do with the problem of missing matter in the universe. The properties of the top quark, which has finally been seen at FermiLab, will be determined. Finally, with the construction of the LHC at CERN (or if we are very lucky perhaps before), we will discover the Higgs particle (or particles) responsible for the electro-weak symmetry breaking—an essential component of electroweak theory which must be experimentally explored.

## 3.2 Beyond the Standard Model

There are two types of attempts to go beyond the standard model. The first are *extensions of the standard model*, by which I mean quantum field theories that incorporate new features, or new symmetries or new dynamics, that go beyond the standard model and explain some of the why questions. The second attempt is to go beyond quantum field theory. The first kind of extension has been contemplated and tried since the beginning of the standard model. Some extensions have already been ruled out or are in deep trouble. For example,

Technicolor, an appealing attempt to provide a dynamical explanation of the electro-weak symmetry breaking is essentially ruled out by experiment.

The traditional way of answering why questions in elementary particle physics has been to push experimental exploration to shorter distances using higher energy accelerators. This path has been followed. Over the last decades the energy of accelerators has increased by more than an order of magnitude. Beautiful experiments have been carried out that look for deviations from the standard model that would be an indication of new physics—but to no avail. All experiments are in accord, with ever increasing accuracy, with the predictions of the standard model, and no surprises have been found. Theorists on the other hand, using pencil and paper, have extrapolated even farther. Shortly after its formulation, when it was realized that the forces of nature vary in strength with energy, the standard model was extrapolated to very high energies.

As the energy increases the electromagnetic force grows stronger and the nuclear force grows weaker. The latter phenomena is *asymptotic freedom*, which explains why nuclei appear to be made of almost non-interacting quarks at high energy, whereas they are strongly bound at low energy. The early extrapolations indicated that the three gauge couplings of the standard model converged to the same value at an energy of roughly $10^{14}$ GEV. Since the gauge forces are so fundamental to the standard model, this extrapolation suggested that the next relevant threshold at which further unification took place was at this very high energy. Indeed the families of quarks and leptons that have been observed fit neatly into patterns dictated by a unified symmetry group, a *grand unified theory (GUT)*, that could be operative at such energies and then split into the various forces seen at lower energies.

Such a picture of grand unification makes the hierarchy problem, the enormous ratio of the GUT scale to the low energy scale even more severe. One beautiful solution to this problem is related to a new symmetry contemplated by theorists—*supersymmetry*. Supersymmetry is one of the most attractive ideas to emerge from theoretical speculation. Why do theorists love supersymmetry?

- First, there is its beauty. It is a elegant extension of ordinary spacetime symmetries. In a nutshell we can understand supersymmetry to be the symmetry of a spacetime which has new dimensions, whose coordinates are not ordinary numbers but *anti-commuting numbers*, so that if $\theta_1$ and $\theta_2$ are two such coordinates then $\theta_1\theta_2 = -\theta_2\theta_1$. This symmetry has the potential of explaining why fermions (particles with half integer spin such as the quarks and leptons) exist. Supersymmetry predicts that for each particle that we have hitherto observed there exists a corresponding su-

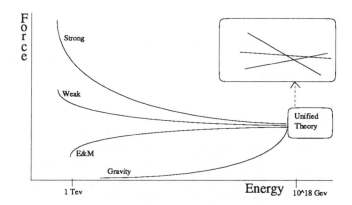

Figure 1: The Forces of Nature as a Function of Energy.

perpartner whose intrinsic spin differs by one-half. Since no such particles have yet been observed supersymmetry must be spontaneously broken. Nonetheless, the existence of the symmetry can put forces (mediated by gauge mesons) and fermionic matter on an equal footing.

- Second, supersymmetry is an automatic consequence of string theory. Indeed supersymmetry was first noticed in string theory, because in string theory it is essentially necessary to have supersymmetry if the theory is to contain fermions at all.

- Most important, supersymmetry offers the most plausible solution that we know to the hierarchy problem. In this attractive scenario the stability of the electro-weak gauge symmetry at the unification scale is protected by supersymmetry, whose breaking then drives symmetry breaking at a much lower scale. The large top quark mass, for which we have no explanation, in fact offers a natural explanation of the generation of the electroweak scale since quantum corrections from virtual top quarks give rise to a mechanism that drives the instability and produces at the electroweak scale the spontaneous symmetry breaking that generates masses for the $W$ and $Z$ particles and for the fermions. For this solution to work we must have low energy supersymmetry, the masses of the super partners must be of order $100$ GeV$-1$TeV, precisely in the regime accessible by the next generation of accelerators (LHC), or if we are lucky by current accelerators.

- Finally, many supersymmetric models contain a particle (the lightest supersymmetric partner) that has the properties to be a natural candidate for the missing matter in the universe.

There is already indirect evidence for supersymmetry. The high precision measurements of the standard model parameters can be used to improve the extrapolation of the gauge couplings to very high energy. These extrapolations contradict the simple predictions of minimal GUT's without supersymmetry, yet they are in beautiful agreement with the simplest predictions of minimal supersymmetric unification. In this model all the couplings coincide at an energy of $\approx 2.10^{16} GeV$. The increase in the GUT scale is due to the effects on the strong coupling of the extra supersymmetric matter. This agreement cannot be taken as proof of supersymmetry, but if it had not worked it would have been taken as good evidence against minimal supersymmetry.

Experimenters now engaged in planning the detectors and experiments for the new accelerators are focusing on detecting supersymmetric partners of ordinary matter. We eagerly await the experimental discovery of the signs of this (spontaneously broken) symmetry at the next generation of particle accelerators. This will be an experimental discovery of the highest importance, in a sense the discovery of extra (anti-commuting) dimensions of spacetime. The discovery of supersymmetry (or strong evidence for its absence) would open up one of the golden ages of experimental physics and provide us with essential clues to the the the unification of the forces.

## 4  String Theory

During the last decade there have been attempts to go significantly beyond the standard model. These attempts have included such new ideas such as supergravity and unification of gauge interactions with gravity by considering theories in more than four dimensions. The last two are examples of ideas that have failed since they were unable to reproduce the standard model.

What we clearly need is a theory that unifies the forces at $\sim 10^{17} GeV$, that contains quantum gravity and that could be able to produce the standard model gauge group at low energy and the observed families of quarks and leptons. This theory should have no adjustable parameters so that we could explain all the parameters of the standard model. Where are we going to find such a theory? Well, there happens to be one, a theory that was accidentally discovered for other reasons and developed over the last 25 years. This theory is string theory.

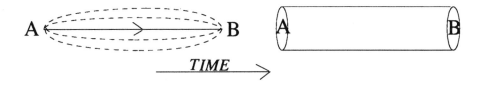

Figure 2: The Propagation of Particles and Strings

We used to think that the proton was an elementary pointlike particle and then, we learned that at distances of a Fermi (or $10^{-13}$ cm.), it has structure, it is made out of quarks. At present energies we can only explore distances down to $10^{-17}$ cm., and the quarks look pointlike. Many people have wondered whether when we look at shorter distances, we will not see that each quark is made out of subquarks. But history does not always repeat itself. String theory says that if we look at a quark with a good microscope, that can resolve distances of $10^{-33}$ centimeters, we will not see smaller constituents, but rather the quarks will look to us like a little closed string. Instead of being pointlike they are extended, one-dimensional objects.

I cannot explain here how string theory works in detail, but I would like to emphasize that the way we have constructed string theory is a natural generalization of the way we construct theories of particles. For example, in classical physics particles move, as time evolves, along trajectories of minimal length. In other words, of all possible motions the actual motion is the one for which the path traversed has the smallest possible length. In flat space a particle, if there are no other particles around, will therefore move in a straight line. The dynamics of strings is constructed by generalizing this same principle to extended objects. We say that strings as well, as they evolve in time, move along a trajectory in such a way that the area of the tube they span is as small as possible. Based on that principle one can construct both the classical and the quantum mechanical description of the propagation of strings. Unlike point particle theories we do not now possess a more fundamental principle on which to base string theory. All we have are adhoc rules for constructing the probability amplitudes for the propagation of strings in a semi-classical expansion about some background. Nonetheless, this has already produced some amazing consequences.

At first people studied the modes of vibration of both closed strings and open strings and looked for their properties, i.e. the masses and quantum numbers of the natural vibrations of these strings. The remarkable thing that they discovered was that closed strings always contained a particle that could be identified with the graviton, the quantum of gravity, and that open strings always contained a particle that could be identified with gauge mesons. This came out of the theory without having to be put in by hand. In fact, it was very embarrassing because originally string theory was constructed as a theory of nuclear force. As such there was no room for gravity or electromagnetism. It is only with the revival of string theory in the 1980's, as a unified theory of everything, that this feature is very welcome. The other remarkable, and originally embarrassing, feature of string theory was that these theories were only consistent if one imagined that spacetime was 26 dimensional (For the more realistic supersymmetric superstrings, the dimension of spacetime has to be ten). Again, as a theory of the nuclear force this is absurd, but it is quite tolerable in the context of a unified theory containing gravity in which the global structure of spacetime is a dynamical issue and extra spatial dimensions could be curled up into small manifolds and thus not directly observable.

The biggest difference between particles and strings appears when we come to interactions, to the forces that exist between particles or strings. We can think about interactions between particles in terms of the trajectories that describe their motion by saying that when two particles (say A and B) meet at the same point they have some probability of turning into a third particle (C), and then that third particle with some probability can turn into two particles (D and E). Thus we have a scattering process, where particle A scatters off particle B to produce particles D and E. The interaction, therefore, is all concentrated at the point where the trajectories meet, a singular point of the diagram, or graph, that describes their spacetime evolution. The introduction of such an interaction at a point is an ad hoc and highly non-unique procedure, which is one of the reasons there are so many particle theories in the world.

How do strings interact? We would like to let strings interact locally as well by having two strings come together and when they touch at a point become a third string. We describe this by the so-called pants diagram. Think of horizontal slices through your pants, and you will see that this describes the time history of two strings, coming together and forming a third string. However there is no particular point, no singular point, where the strings join together. Unlike the particle picture there is no point that you can pick out and say "this is where the interaction took place." The surface is completely smooth. It is essentially because of this natural geometrical and unique way of introducing interactions that strings are so symmetrical and unique.

The uniqueness of string dynamics, in which the form of the interaction is totally fixed, is suggestive of a very large symmetry of the theory that dictates the interaction. String theory certainly possesses an incredibly large symmetry.

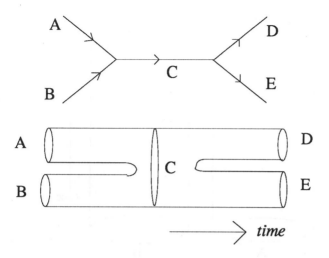

Figure 3: The Interaction of Particles and Strings

What is this marvelous symmetry of string theory? We do not know. It is one of the main concerns of current research to discover the full symmetry of string theory. The full symmetry is incredibly large, but it is hard for us to see because it is largely broken or hidden from us. What we do know is that string theory contains automatically, without our arranging for it ahead of time or adjusting anything, the largest symmetry that has ever been conceived by point particle physicists. It contains automatically the symmetries that are responsible for the emergence of gravity and the other gauge interactions of nature.

   Still in its infancy, string theory already has remarkable achievements. It provides:

- a consistent, logical and rich extension of the conceptual structure of physics;

- a consistent and finite theory of quantum gravity;

- a rich structure which might yield a unique and comprehensive description of the real world.

There have been two major revolutions completed in this century: relativity, special and general, and quantum mechanics. These were associated with two of the three dimensional parameters of physics: $\hbar$, Planck's quantum of

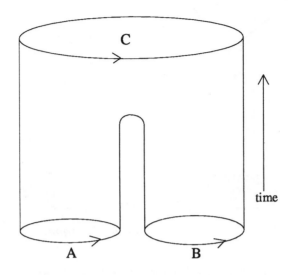

Figure 4: The *Pants Diagram* That Describes String Splitting

action, and $c$, the velocity of light. Both involved major conceptual changes in the framework of physics, but reduced to classical non-relativistic physics when $\hbar$ or $1/c$ could be regarded as small. The last dimensional parameter is Newton's gravitational constant which sets the fundamental (Planck) scale of length or energy. Some of us believe that string theory is the revolution associated with this last of the dimensional parameters of nature. At large distances compared to the string scale of $10^{-33}$ cm., string theory goes over into field theory. Meanwhile string theory does provide us for the first time with a finite and consistent theory of quantum gravity, and thus an existence proof that gravity and quantum mechanics are mutually consistent. Many people suspected that this was not the case, so it is important to have such an example. In addition, stringy gravity is a very interesting theoretical tool. It has motivated the construction of many low energy models of (super)gravity. It has also provided some soluble toy models of two dimensional gravity. These provide a very useful theoretical laboratory for discussing interesting and important conceptual issues of quantum gravity, especially those related to the fate of black holes.

Finally, string theory does provide us with a rich structure which could yield the real world. We have discovered that string theory automatically contains within it all of the ingredients that we seem to need to explain the standard model, and that in principle it contains no arbitrary parameters. Thus it offers the possibility of a truly unified theory of everything—that is everything that we know of at present. Perhaps the final theory; perhaps the next theory that will lead to new why questions.

The reason we are unable to construct predictive models at the moment is the lack of understanding of the nonperturbative dynamics of string theory. In our eventual understanding of string theory we might have to undergo a discontinuous conceptual change in the way we look at the world similar to that which occurred in the development of relativity and quantum mechanics. I think that we are in some sense in a situation analogous to where physics was in the the beginning of the development of quantum mechanics, after Bohr's model of the atom, where physicists only had a set of ad-hoc rules. These turned out in time to be the semiclassical approximation to quantum mechanics, but originally were not part of a consistent, coherent framework. There was an enormous confusion, until in the mid 1920's quantum mechanics was finally discovered. It can be argued that the final stages of this development were motivated primarily by theoretical considerations, the necessity for making sense out of the contradictions and paradoxes of the "old quantum mechanics."

We might now be in a similar situation, and eventually will be required to make a revolutionary break with previous concepts and develop the "new string theory." Where will this revolution lead? Which of our concepts will have to be modified? There are many hints that our concepts of spacetime, which are so fundamental to our understanding of nature, will have to be altered.

The first hint is based on a stringy analysis of the measurement of position, following Heisenberg's famous analysis in quantum mechanics. Already in ordinary quantum mechanics space becomes somewhat fuzzy. The very act of measurement of the position of a particle can change its position. In order to perform a measurement of position $x$, with a small accuracy, $\Delta x$, we require probes of very high energy $E$. That is why we employ microscopes with high frequency (energy) rays or particle accelerators to explore short distances. The precise relation is that

$$\Delta x \approx \frac{\hbar c}{E},$$

where $\hbar$ is Planck's quantum of action and $c$ is the velocity of light. In string theory, however, the probes themselves are not pointlike, but rather extended objects, and thus there is another limitation as to how precisely we can measure short distances. As energy is pumped into the string it expands and thus there is an additional uncertainty proportional to the energy. All together

$$\Delta x \approx \frac{\hbar c}{E} + \frac{GE}{c^5},$$

where $G$ is Newton's gravitational constant. Consequently it appears impossible to measure distances shorter than the Planck length.

The second hint is based on a symmetry of string theory known as duality. Imagine a string that lives in a world in which one of the spatial dimensions is a little circle of radius $R$. Such situations are common in string theory and indeed necessary if we are to reconcile the fact that the string theories are naturally formulated in nine spatial dimensions so that if they are to look like the real world, six dimensions must be curled up, *compactified*, into a small space. Such perturbative solutions of realistic string theories have been found and are the basis for phenomenological string models. Applied to the simple example of a circle of radius $R$, duality states that the theory is identical in all of its physical properties to one that is compactified on a different circle of radius $\bar{R} = L_p^2/R$, where $L_p$ is the ubiquitous Planck length of $10^{-33}$ cm. Thus if we try to make the extent of one of the dimensions of space very small, by curling up one dimension into a circle of very small radius $R$, we would more naturally interpret this as a world in which the circle had a very large radius $\bar{R}$. The minimal size of the circle is of order $L_p$! This property is inherently stringy. It arises from the existence of string states that wind around the spatial circle and again suggests that spatial dimensions less then the Planck length have no meaning.

Finally, in string theory the very topology of spacetime can continuously be altered. In perturbative string theory there are families of solutions labeled by various parameters. In some cases these solutions can be pictured as describing strings propagating on a certain curved spatial manifold. As one varies the parameters the shape and geometry of the background manifold varies. It turns out that by varying these parameters one can continuously deform the theory so that the essential geometry of the background manifold changes. Thus one can go smoothly from a string moving in one geometry to a string moving in another; although in between there is no simple spacetime description.

All of these hints suggest that spacetime in string theory is not a fundamental property, but rather an approximate way of describing the physics in certain domains. But what replaces spacetime is still totally mysterious.

## 5   What Will the Future Bring?

The rumors of the death of particle physics have been greatly exaggerated. Building on the grand synthesis of the standard model, we are in a position to address some of the deepest issues in fundamental physics. Existing accelerators and new machines that are now under construction should allow us to extend our understanding of standard model physics, and to confirm the mechanism of symmetry breaking that produces masses for the elementary constituents of matter. Most exciting is the possibility that these instruments can discover whether the world is supersymmetric or not, providing us with crucial hints as to the next stage of unification.

On the theoretical side we have the beginnings of a new theory of fundamental physics—string theory, whose full elucidation could be as revolutionary as the discovery of quantum mechanics.

Talented young people should not now be discouraged by gloomy claims about the state of basic physics. The problems are great, but the opportunities are great also. Particle physicists are grappling with wonderful questions and marvelous, mysterious ideas. There is every possibility of new synthesis in coming years, and what looks like calm might be the eye of the storm.

# STATUS AND PROSPECTS OF
# RELATIVISTIC HEAVY-ION EXPERIMENTS

## S. NAGAMIYA*

*Department of Physics, Columbia University*
*538 W120th St., New York, NY 10027, U.S.A.*

In this article I would like to describe a) major goals in the field of relativistic heavy-ion physics, b) three most recent data, and c) future prospects in particular plans at Relativistic Heavy-Ion Collider (RHIC) at Brookhaven National Laboratory.

## 1. Physics of Relativistic Heavy-Ion Collisions

*1.1. Hadron Identity in Nuclear Matter*

Recently, nuclear and particle physicists asked almost the same question from two different angles. Nuclear physicists' question is as follows: The atomic nucleus on the Earth has a constant density of 0.17 nucleons/fm$^3$ ($\approx$ 0.3 billions tons per cc), regardless of the species of nuclide. This fact creates two questions in nuclear physics, as illustrated in Fig 1. One is how the energy of nuclear matter changes as a function of nuclear matter density, $\rho$. This is known as a question of "Equation of State".

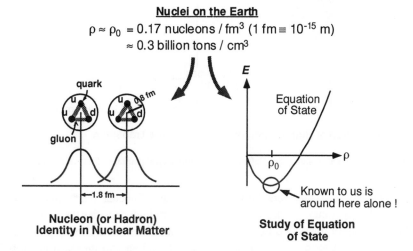

**Nuclei on the Earth**

$\rho \approx \rho_0$ = 0.17 nucleons / fm$^3$ (1 fm $\equiv$ 10$^{-15}$ m)
$\approx$ 0.3 billion tons / cm$^3$

*E*

quark

0.8 fm

gluon

Equation of State

$\rho_0$

$\rho$

←1.8 fm→

Known to us is around here alone !

**Nucleon (or Hadron)**
**Identity in Nuclear Matter**

**Study of Equation**
**of State**

Figure 1: The constant density of atomic nuclei leads to two nuclear physics questions.

---

* Present address: Institute for Nuclear Studies, University of Tokyo, Midori-cho, Tanashi-shi, 188, Japan.

Known to us is at $\rho \approx \rho_0$ alone. Studying other regions is a challenging task in nuclear physics. The other question is on the identity of nucleon or, in general, that of hadron in nuclear matter. The constant density of the nucleus means that the inter-nucleon distance ($d$) inside the nucleus is nearly constant ($d \approx 1.8$ fm). On the other hand, as shown in Fig. 1, the nucleon has a finite radius of $\sqrt{<r^2>} \approx 0.8$ fm with its charge distribution extending beyond its radius, because the nucleon is made of more fundamental particles: quarks and gluons. This radius is only one half of the inter-nucleon distance. Therefore, once a nucleon is imbedded inside the nucleus, the nucleon might have a certain overlap with neighboring nucleons and, as a result, the wave function of that nucleon might be distorted and be different from that of a "free" nucleon. If the nucleon loses partially its identity in nuclear matter, then the nucleon's sub-degrees of freedom (quarks and gluons) would play a role in the behavior of the nucleon inside the nucleus.

For example, at $\rho = 5\rho_0$ the neighboring nucleons would overlap significantly like the case as shown in Fig. 2. In this case, the identity of the nucleon might be lost significantly, so that the system melts into a soup of quarks. Accordingly, Equation of State would also change significantly, as shown in Fig. 2. Studying of Equation of State at high density is, thus, related directly to the question of how the property of hadron changes in nuclear matter.

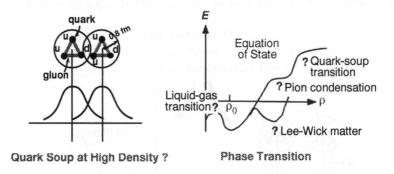

Figure 2: Nuclear matter at $5\rho_0$ (left) and speculations on Equation of State of the nucleus.

Nuclear physicists therefore drew a phase diagram of nuclear matter in the plane of density and temperature, as shown in Fig. 3. At very high densities the system would melt into a soup of quarks. Similarly, if nuclear matter is heated, many pions would be created. Because the pion is made of a quark and an anti-quark, it has a finite radius of $\sqrt{<r^2>} \approx 0.6$ fm. At sufficiently high temperatures a substantial overlap among pions would be expected and the system would melt into a soup of quarks and anti-quarks. The study of nuclear matter at high densities and/or at high temperatures is, thus, very intriguing to answer the question at which conditions the hadrons lose completely their identity.

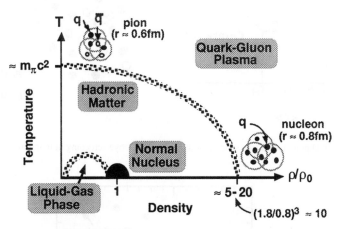

$T \approx m_\pi c^2 = 140$ MeV: Hagedorn's limiting temperature

Figure 3: Expected phase diagram of nuclear matter.

## 1.2. Quark Confinement and Deconfinement

Particle physicists, on the other hand, has asked a question why quarks are confined. Analytically, the confinement mechanism is not easy to understand, so that computer calculations, called the lattice QCD calculation, have been introduced.

Phenomenologically, quark-antiquark (or quark-quark) interactions have the property such that they becomes stronger as the inter-quark distance increases. This spring-like potential, which is schematically illustrated by the solid curve in Fig. 4, explains intuitively the mechanism of quark confinement. The meson or, in general, hadron is defined as a bound state of this potential. Therefore, the first task of lattice QCD calculations is to determine the meson mass in free pace.

In parallel to the effort to calculate meson masses, lattice QCD theorists asked another important question: the question at which conditions the confinement disappears. For this purpose, a hot QCD vacuum has been studied. Results of these calculations are such that at high temperatures (of the order of 150-200 MeV) the confinement disappears completely. A typical result is shown in Fig. 5.[1] If the confinement disappears in a hot vacuum, then the system would melt into a soup of quarks, anti-quarks, and gluons. This matter is called the quark-gluon plasma.

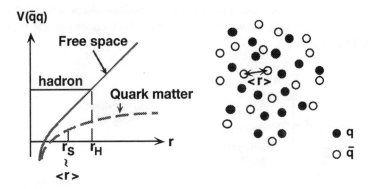

Figure 4: QCD potential in free space (solid curve) and Debye screening at high temperatures.

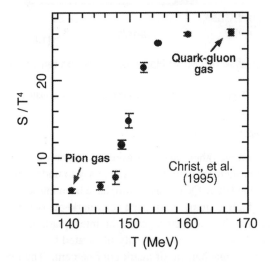

Figure 5: A recent result of lattice QCD calculations to show a formation of quark-gluon plasma at high temperatures. Result taken from Ref. 1.

An analogy of deconfinement mechanism in daily life would be as follows: If I speak with Mr. X alone, then our mutual interaction would be strong. However, if I speak with him in the presence of a large number of people, then our mutual interaction will be screened and weakened by other people. This screening effect would be stronger as our mutual distance increases. A similar situation occurs in the QCD potential. The screening length, $r_D$ (called the Debye length), is on the order of inter-quark distance.

As the temperature increases, the number of quarks and anti-quarks increases, and the Debye screening length decreases. At sufficiently high temperatures, where the screening length becomes shorter than the hadron radius itself, like the case as shown by the dashed curve in Fig. 4, then this screened potential can no longer create any bound states and each quark moves almost freely inside the system. This is the reason that the system is called a deconfined plasma. Clearly, the deconfinement is a many-body effect.

The present hot topic in the lattice QCD calculations is to study if the phase transition is the first order or the second order. Fig. 6 illustrates the order of phase transition plotted as a function of quark masses. The real world is indicated by the closed

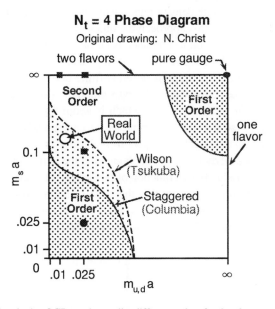

Figure 6: Two lattice QCD results predict different orders for the phase transition.

circle. Two groups used different methods to calculate: The Columbia group[1] used a staggered method and concluded that the phase transition is the second order, whereas the Tsukuba group[2] used the Wilson loop and concluded that it is the first order. This difference is due to a limited computer speeds. Both methods must give the same answer with an ideal computer. In the near future, when a Telaflop-type computer becomes available, this question would be solved.

The other element, which occurs at the phase transition, is the change of chiral symmetry which induces a reduction of effective quark mass. Although I skip to describe details, it is known that the chiral symmetry of the QCD vacuum is broken ($<\bar{\psi}\psi> \neq 0$) and this symmetry breaking creates the mass of a hadron. According to lattice QCD calculations, the quantity $<\bar{\psi}\psi>$ approaches zero when the phase transition occurs. This phenomenon is called the chiral symmetry restoration. Accordingly, the effective quark mass at above the transition temperature becomes almost zero for $u$ and $d$ quarks and 150 MeV/$c^2$ for the $s$ quark. The situation is very similar to an insulator-conductor phase transition in solids, where effective electron mass changes in association with the phase transition.

So far, lattice QCD calculations were performed only for the zero-density limit. However, efforts are in progress to extend these calculations into the high baryon density region. For the finite density region a different theoretical approach was also developed.[3] According to this approach the quantity $<\bar{\psi}\psi>$ was evaluated based on an analytical method based on a spontaneous chiral symmetry breaking proposed by Nambu and Jona-Lasinio.[4] The most striking result is that the chiral symmetry breaking of the vacuum is partially restored at finite baryon density. Fig. 7 shows a typical result. The expectation value of $<\bar{\psi}\psi>$ gradually decreases as the matter density increases. At the critical density (which is about $5\rho_0$ in this calculation), this value becomes zero, so that the system melts into a soup of quarks. An important point is that, although the quantity $<\bar{\psi}\psi>$ drops sharply as a function of temperature in this calculation, similarly to the result obtained in the lattice QCD calculations, this quantity drops only gradually as a function of the density of nuclear matter.

This calculation further implies that the mass of baryon or meson is reduced in nuclear matter at finite density. Efforts to measure meson masses in nuclear matter have, thus, began recently. This point will be described briefly in Section 2.2.

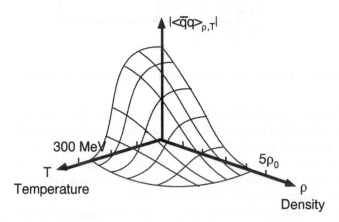

Figure 7: Hadrons in nuclear matter. Figure taken from Ref. 3.

## 1.3. Hot/Dense Matter and Heavy-Ion Collisions

Can experimentalists create and study hot and dense matter in heavy-ion collisions? Nuclear collisions at high energies can be described as two clouds of nucleons colliding with each other and suffering many sequential nucleon-nucleon interactions. The situation is somewhat like an ancient battle. Suppose I am a warrior about to engage in hand-to-hand combat. When the battle starts, my opponents press toward me, while at the same time, my compatriots close ranks from behind me. Therefore, the local density around me suddenly increases. Furthermore, during the battle everyone's available energy is distributed among the many combatants, which is called thermalization. For example, if I attempt to stand still, I am attacked from different directions. By defending myself I transfer energy to many others. The overall temperature on the battlefield sharply increases. Then, as the conflict subsides, the survivors (including me, I hope) become exhausted and eventually disperse. Similarly, in a heavy-ion collision the local density increases, the nucleons undergo thermalization, the temperature rises rapidly, and then both density and temperature decrease after the collision. Therefore, heavy-ion collisions allow investigations at both high densities and high-temperatures.

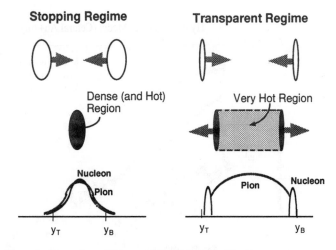

Figure 8: Heavy-ion collisions and the formation of hot/dense matter.

Theorists expect that at beam energy of 10–100 A•GeV, which is the energy region covered by the AGS accelerator at Brookhaven National Laboratory (BNL) and the SPS accelerator at CERN, there is a possibility for the system to sweep a region of high matter density. At RHIC (the Relativistic Heavy-Ion Collider), which is under

124

construction at Brookhaven National Laboratory (BNL), the energy is 100 A•GeV in a collider mode. Colliding nuclei will penetrate through each other, possibly creating bunched gluon strings in the central region immediately following the collision. Fig. 8 illustrates how a very hot vacuum would be created over an extended volume at RHIC. While RHIC is expected to become operational in 1999, several years later such experiments will be extended at the forthcoming LHC (Large Hadron Collider) at CERN. At the LHC a baryon free hot vacuum could also be created.

## 2. Recent Heavy-Ion Data from the BNL-AGS and the CERN-SPS

### 2.1. Hadron Spectra and Compression/Explosion

Fig. 9 shows the energy spectra of pions, kaons and protons produced in the Au + Au collisions at 11.6 A•GeV/c. The most central collision was selected by filtering with the highest charged-particle multiplicity produced in the collision. The spectra are those for particles emitted at $90^\circ$ in the center-of-mass frame.[5] Notice that the slope for pions is steeper than that for kaons, and the slope for kaons is steeper than that for protons. Thus, we observe that the slope is generally steeper for lighter-mass parti-

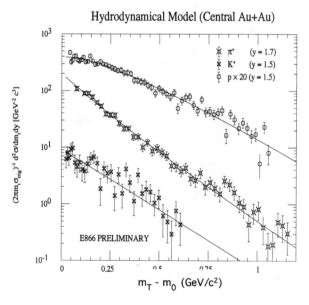

Figure 9: Spectra of p, $\pi$ and K measured in 11.6 A•GeV/c Au + Au collisions at the AGS.

cles. If the particles were emitted from a thermal source, then the spectrum would show a pure exponential shape in the energy spectrum with the same slope for all particle species. Note also that the spectra in Fig. 9 are not purely exponential.

These spectral trends were known for many years since the time when the first data from the Bevalac were published.[6] Fig. 10 illustrates an explanation. If the nucleus is compressed, then the system would explode. Assume that the particle spectrum is a superposition of a chaotic thermal-like spectrum and an explosion flow. For the chaotic part the spectrum would be a simple exponential with the same slope for all the particles. However, because the velocity of lighter-mass particle is larger than that of heavier-mass particle for a give kinetic energy, the influence of the explosion flow is stronger for a slower (and, thus, for a heavier-mass) particle than for a lighter-mass particle. In other words, the apparent slope would be less steeper for heavier-mass particles than for lighter-mass particles. In addition, the effect of the flow is significant more in the low-energy region than in the high-energy region in the spectrum. The solid curves in Fig. 9 show that fitting the data using the above idea works very well. Here, the parameters are $v_{flow}, = \beta_{flow}c$ and $T_0$.

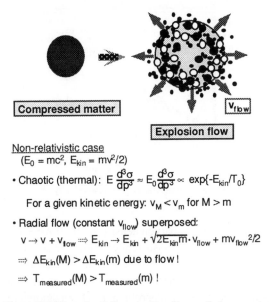

Non-relativistic case
($E_0 = mc^2$, $E_{kin} = mv^2/2$)

• Chaotic (thermal): $E\frac{d^3\sigma}{dp^3} \approx E_0\frac{d^3\sigma}{dp^3} \propto \exp\{-E_{kin}/T_0\}$

For a given kinetic energy: $v_M < v_m$ for $M > m$

• Radial flow (constant $v_{flow}$) superposed:

$v \to v + v_{flow} \Rightarrow E_{kin} \to E_{kin} + \sqrt{2E_{kin}m}\cdot v_{flow} + mv_{flow}^2/2$

$\Rightarrow \Delta E_{kin}(M) > \Delta E_{kin}(m)$ due to flow !

$\Rightarrow T_{measured}(M) > T_{measured}(m)$ !

Figure 10: Effect of radially expanding flow on hadron spectra.

By studying the data on particle production in heavy-ion collisions at the Bevalac (Berkeley), SIS (GSI in Germany), AGS (Brookhaven) and SPS (CERN), one can

126

make a systematic study on the variation of $\beta_{flow}$ and $T_0$ as a function of beam energy. Fig. 11 shows the summary of those values presented by Stachel[7] at the Quark Matter Conference in 1996. One learns that, although the value of $T_0$ increases monotonically with the beam energy, the value of $\beta_{flow}$ reaches its maximum value at around 10-30 A•GeV beam energies. This feature is striking. It would mean that the nucleus is compressed most effectively (and, thus, explodes most strongly) at these energies, so that the formation of the high density matter is most efficient at 10-30 A•GeV beam energies.

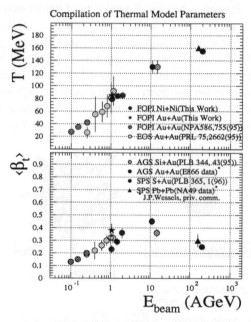

Figure 11: Beam-energy dependence of temperature and flow velocity. Figure provided by Stachel.[7]

The formation of the maximum baryon density at 10-30 A•GeV has been expected also from earlier studies of nuclear transparency. However, a systematic study on the explosive flow made it possible for the first time to indicate clearly that the maximum compression occurs in this beam energy region. What are the maximum densities attainable in heavy-ion collisions? This problem has been studied first by Landau[8] in 1953. If two objects collide each other and stop completely in the center-of-mass frame, then the density to be reached will be:

$$\rho = 2\gamma_{cm} \times \rho_0 \qquad (1)$$

where $\gamma_{cm}$ is the $\gamma$-factor that is used for Lorentz transformation and $\rho_0$ is the density of normal nuclei. The factor 2 comes from the overlap of two colliding nuclei and $\gamma_{cm}$ from the Lorentz contraction of these nuclei. The AGS energy ($\approx 15$ A•GeV/c) leads to $\gamma_{cm} \approx 2.9$ so that we would expect a high density matter at $\rho \approx 6\rho_0$. In fact, recent cascade calculations[9] show that the formation of $(3-4)\rho_0$ could be expected in Si + Au collisions and up to $10\rho_0$ in Au + Au collisions.

## 2.2. ρ-Meson in Dense Nuclear Matter

Encouraged by a hint on the formation of high density nuclear matter in heavy-ion collisions at the AGS and SPS beam energies, efforts were begun to measure the masses of vector mesons in nuclear matter in these collisions. The first trial of this type was to measure the mass of $\phi$-meson in Si + Au collisions at the AGS from the measurement of $\phi \rightarrow K^+K^-$.[10] In this case, however, no significant mass shifts were observed. Clearly, it is not suitable to measure any meson masses via hadronic decay channels, since hadrons interact strongly with nuclear matter and wash out the memory of the initial information on the meson mass. Lepton pairs must be measured to determine whether the meson mass in nuclear matter is different from that in free space.

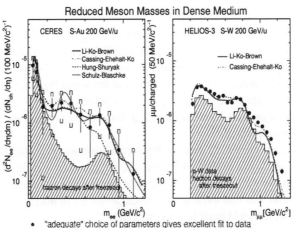

Figure 12: Dielectron production in heavy-ion collisions, as compared with the corresponding data (hatched region) taked in p+Be collisions. Solid curves are the fit of the data with a reduced ρ-meson mass.

Recently, dielectron measurements were performed at the CERN-SPS. The NA45 experiment[11] showed that the data of dielectron mass spectrum in p + Be collisions at 450 GeV beam energy can be fit perfectly with the known decay contributions from various mesons in free space. However, in S + Au or Pb + Au collisions an enhanced yield above the superposition of the known decay contributions was observed, as shown in Fig. 12. In particular, considerable enhancement is observed in the invariant mass region of $\approx 0.4$ GeV/c$^2$.

The free $\rho$-meson has a mass of 0.77 GeV/c$^2$. One of the most plausible theoretical models to explain the data is based on what was already described in Section 1.2. Namely, the partial restoration of chiral symmetry causes a meson mass shift in high density nuclear matter. In fact, Fig. 12 shows that the data can be explained reasonably well with this idea. On average, the $\rho$-meson mass could be shifted by 40-50% in these calculations.

Of course, the study of meson masses has just begun and it has to be continued to pin down the question if the hadron mass actually changes in nuclear matter. Experimental data need higher statistics. Theoretical models need much more improvement. However, heavy-ion beams at the AGS and SPS will be closed soon. Discussions of heavy-ion accelerators in the energy domain of 10-30 A•GeV region are, thus, in progress in both Japan and Germany. I personally feel that such studies must be continued even in the next century.

### 2.3. J/ψ Suppression

For many years a systematic measurement of the yields of $J/\psi$ and $\psi'$ via leptonic channels has been carried out at the CERN-SPS to study the Debye screening length of the QCD potential. As described in Section 1, this Debye screening induces deconfinement. Therefore, the measurement of the screening length is important. The $J/\psi$ is a bound state of $\overline{c}c$ but, if $r_D < r(J/\psi)$, which is likely to happen, the QCD potential cannot create a bound state of $J/\psi$ in the phase of QGP and, thus, this meson production is suppressed.[12]

Recently, an interesting data on $J/\psi$ suppression was reported from NA50 at the CERN SPS, as shown in Fig. 13.[13] The $J/\psi$ suppression was already observed in p + A collisions as well as S + A collisions in earlier data. A standard explanation of this suppression was that the $J/\psi$ meson was absorbed in nuclear matter after its production. Thus, the yield must be proportional to[14]

$$N(J/\psi) \propto \exp(-\rho L \sigma_{abs}) , \qquad (2)$$

where $\rho$ is the matter density, $L$ is the distance through which the $J/\psi$ traverses in nuclear matter and $\sigma_{abs}$ is the absorption cross section of the $J/\psi$. The straight line in

Fig. 13 is a fit with $\sigma_{abs}$ = 6.2 mb. For Pb + Pb collisions the $J/\psi$ yield is significantly smaller than the extrapolation of this straight line fit. In order to explain this anomaly, models such as the co-mover model, which assumes that the $J/\psi$ absorption is proportional to the total number of particles produced in the collision instead of the length, $L$, have been introduced. A full explanation of the data, however, has not been settled.

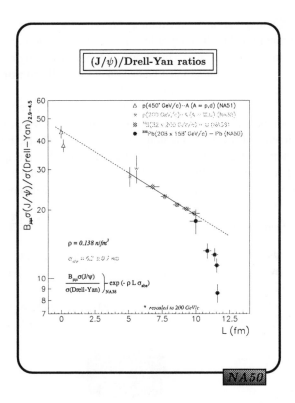

Figure 13: Systematics of the $J/\psi$ yield measured by NA50 in p+A and A+A collisions.

# 3. Physics at RHIC

## 3.1. RHIC, Particle Multiplicity and Energy Density

The most exciting topic at present in this field is an expected operation of the Relativistic Heavy-Ion Collider (RHIC) from 1999. RHIC will provide Au beams at

100 A•GeV in a collider mode at luminosity of $2 \cdot 10^{26}/cm^2 \cdot s$. Later, polarized proton beams will also be available.

In central Au + Au collisions at RHIC, the expected charged-particle multiplicity is $\approx$ 5,000. The density of this multiplicity reaches $dN_{ch}/dy \approx 1,000$ at $y = y_{cm} = 0$, where $y$ is the variable called the rapidity defined as

$$y \equiv \frac{1}{2}\ln\left(\frac{E + p_L}{E - p_L}\right), \tag{3}$$

where $p_L$ is the longitudinal momentum and $E$ is the total energy of the particle. The rapidity is proportional to the longitudinal momentum in a non-relativistic case, but it is a Lorentz invariant quantity. The above multiplicity density corresponds to $dN_{ch}/d\Omega \approx 150$ at $y = 0$. Detectors must be designed to handle an extremely high multiplicity and, clearly, the task is not trivial.

As described in Section 1, the phase transition is expected at $T_C \approx 150$ MeV. Below this temperature, the system is a gas of pions. In the pion system, internal degrees of freedom ($f$) are isospin alone and, thus, $f = 3$. On the other hand, at above $T_C$, the system is a gas of quarks and gluons. Because the quark carries spin, isospin, and color and the gluon has polarization and color, the total degrees of freedom are $f = 37$. In both cases, the system obeys Stefan-Boltzmann law:

$$\varepsilon_{\text{pion gas}} = 3 \times \frac{\pi^2 T^4}{30}, \qquad \varepsilon_{\text{quark-gluon gas}} = 37 \times \frac{\pi^2 T^4}{30}. \tag{4}$$

This implies that, at $T_C$, the energy density which is required to create QGP is by a factor 12 larger than that for a pion gas If $T_C = 150$ MeV, the required energy density for the formation of QGP is $\varepsilon \approx 1$ GeV/fm$^3$.

In heavy-ion collisions at RHIC the total energy accumulated in a hot vacuum can be estimated by the following method. For notations see Fig. 14. The accumulated

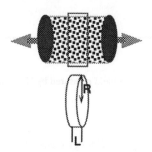

Figure 14: Collision geometry to estimate the energy density.

energy will eventually be emitted by pions. The energy emitted intp a unit rapidity is given by

$$E \approx \frac{3}{2} \frac{dN_{ch}}{dy} \sqrt{m_\pi{}^2 + p_\pi{}^2} \approx \frac{dE_T}{dy} \approx 700 \text{ GeV}. \tag{5}$$

Here, the multiplicity density of $dN_{ch}/dy \approx 1,000$ together with an average pion transverse energy of 0.45 GeV were used. On the other hand, as shown in Fig. 14, the volume of the hot vacuum which is responsible to the emission of particle per unit rapidity is given by

$$V = \pi R_T{}^2 L \approx \pi R_T{}^2 c\tau_0. \tag{6}$$

Note that the space-rapidity relation is given by $dz = c\tau_0 dy$. In Au + Au collisions, $R_T \approx R_{Au}$ (or $R_{Au} + 1$ fm) $\approx$ 6-7 fm. Concerning the value of $c\tau_0$, there is an ambiguity but it is estimated to $c\tau_0 \approx$ 1-2 fm. Therefore, the energy density expected in Au + Au collisions will be[15]

$$\varepsilon \approx \frac{3}{2} \frac{1}{\pi R_T{}^2 c\tau_0} \frac{dE_T}{dy} \approx 3 - 6 \text{ GeV/fm}^3, \tag{7}$$

which is larger than the needed energy density to form QGP. It is likely that QGP is created at RHIC. The question is how to detect QGP.

### 3.2. Signatures for Quark-Gluon Plasma

The primary goal of RHIC experiments is to detect QGP and to study its properties. Many theorists predicted signatures of the QGP, while no predictions are firm, since the heavy-ion collision is a very complicated process and, in addition, the lifetime of QGP which would be created in heavy-ion collisions is short (on the order of $10^{-21}$ s). Probes to detect phenomena which occur for a short time period will be needed. It is my personal belief that, in order to pin down the existence of the QGP, it is not sufficient to measure one predicted signature alone. Simultaneous measurements of several potential signatures of the QGP are important. Furthermore, systematic studies on these signatures as a function of beam energy, as well as these studies as a function of projectile and target masses, will be needed. We will review several potential signatures in what follows.

*J/$\psi$ Suppression* : The first example is a systematic measurement of the yields of $J/\psi$, $\psi'$ and $\Upsilon$ via leptonic channels, to study the Debye screening length of the QCD potential. The suppression of the yield of a vector meson occurs under the condition of $r_D < r(\text{meson})$, where $r_D$ is the Debye screening length. Because $r(\psi') > r(J/\psi) > r(\Upsilon)$, we expect that $\psi'$ melts first in QGP, then, $J/\psi$ melts and, finally $\Upsilon$. According to the present theory, we would not expect any suppression in the yield for $\Upsilon$ in heavy-ion collisions at RHIC. In order to study these yields, measurements of lepton pairs are important at RHIC.

Chiral Symmetry Restoration: As described in Section 1, the quantity of $<\bar{\psi}\psi>$ would be zero in the QGP phase. Accordingly, the mass of quark would approach the bare quark mass. If a certain vector meson can still survive in the form of a bound state of $\bar{q}q$ inside QGP, the mass of this meson would be substantially smaller than that in free space. Measurements of vector meson masses with dileptons at RHIC are, therefore, very important and interesting. For example, we pick an example of $\phi$ meson. The mass of this meson would be changed in the QGP but, at the same time, the mass of K meson would be changed as well. Since the mass of two kaons (= 987 MeV) is very close to the $\phi$ meson mass (= 1,019 MeV), slight changes in K and $\phi$ masses will strongly influence the branching ratio and width of the $\phi$. Note that the main decay mode of the $\phi$ in free space is $\phi \rightarrow KK$. Measurements of these mesons in both leptonic and hadronic decay channels are very important.

Order of Phase Transition: If the phase transition is the first order, then the system will stay for a long period of the time at the critical temperature, $T_C$, until the system completes the release of energy density from the QGP phase to a pion gas phase, because Eq. (4) tells us that the energy density between the two system is different by a factor of $\approx$ 12. This situation is illustrated in Fig. 15. If one projects the upper graph in the plane of $T$ and $\varepsilon$, then, $T$ will have a second rise as a function of $\varepsilon$. Since $T$ is related to $<p_T>$ and $\varepsilon$ to $dE_T/dy$ by Eq. (7), the second rise of $<p_T>$ is expected as a function of $dE_T/dy$. Namely, a typical S-shape phase diagram in the plane of $<p_T>$ and $dE_T/dy$ can be expected.[16] Some years ago, the JACEE cosmic ray group[17] observed a hint of this second rise using heavy-ion beams in cosmic rays at the beam energies at $\sim$ 1 A•TeV. Because RHIC provides a fixed-target equivalent beam energy of $\sim$ 20 A•TeV, it is interesting to study this point at RHIC.

On the other hand, if the phase transition is the second order, then a strong fluctuation could be expected. An interesting example is a fluctuation in the isospin space.[18] In the ferromagnet a large fluctuation of spins at around the Curie temperature is observed in neutron scattering experiments. Similarly, a large fluctuation of isospin could be expected in heavy-ion collisions, as illustrated in Fig. 16. This will induce fluctuations in the ratio of charged to neutral pions. The phenomenon of isospin fluctuation is presently called the disordered chiral condensates (DCC), and the subject is being studied extensively these days.

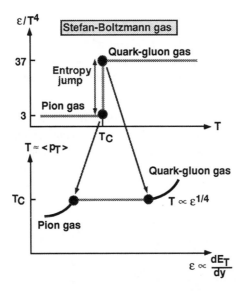

Figure 15: Second rise of <$p_T$> expected in the phase transition.

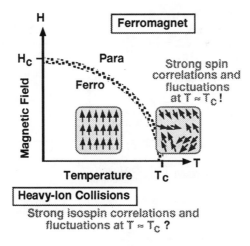

Figure 16: Isospin fluctuation predicted by Wilczek in analogy of magnetic phase transition.

<u>Radiation from Hot Thermal Source, etc.</u>: Theoretically, it has been studied if the intensity of radiation from a quark-gluon gas is significantly different from radiation

intensity from a pion gas. The radiation is caused by bremmstraulung, annihilation of constituents (such as $\bar{q}q \rightarrow \gamma$ for QGP and $\pi^+\pi^- \rightarrow \gamma$ for a pion gas), scattering between constituents (such as $qg \rightarrow \gamma$). Due to the difference in degrees of freedom between the QGP and pion gas, the expected intensity of radiation could differ between these two systems.[19]

Fig. 17 summarizes signatures. As described above, simultaneous measurements of these quantities as a function of a well-defined parameter, which is proportional to $\varepsilon$, would be interesting. This figure shows also typical errors expected from realistic measurements which will be carried out by one of large experiments: PHENIX at RHIC.

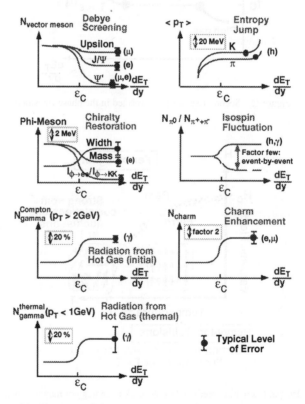

Figure 17: Summary of signatures of quark-gluon plasma phase transition. Expected experimental errors with the PHENIX detector are added.

## 3.3 RHIC Experiments

RHIC has six intersections. Two large experiments, PHENIX and STAR, are under construction. In addition, two small experiments, PHOBOS and BRAHMS, will be installed.

The philosophy of STAR is to try to cover as large solid angles as possible. A TPC will be installed inside a solenoidal magnet to cover the rapidity region of $-2 < y < 2$, to allow tracking of 2,000–3,000 charged tracks. A clear advantage of this approach is that $p_T$ distributions of particles, as well as two-particle correlations, can be measured for one event, because one event contains a few thousand particles. Namely, an event-by-event analysis becomes possible in STAR. Another advantage of this experiment is that a detailed reconstruction of each track is possible so that it will provide a useful tool to measure $V$-particles such as $\Lambda$, $\Xi$, etc. However, a disadvantage is that the capability of particle identification is very limited under the limited amount of the budget. In STAR, $dE/dx$ measurement is the only tool within the approved baseline budget. This method allows K/$\pi$ separation only up to 0.7 GeV/c.

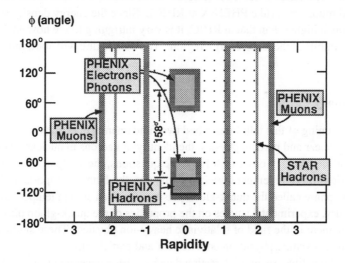

Figure 18: Kinematical coverage by PHENIX and STAR at RHIC..

The other large experiment is called PHENIX which sets a strong emphasis on particle identification by covering a limited solid angle. In the central region a magnetic field will be created parallel to the beam direction by an axial filed magnet. Two-arm spectrometers will be prepared with tracking devices, RICH, TOF and EM calorimeters, to measure electrons, photons and identified hadrons. An event multiplicity will be measured by Si-strip and pad chambers over a wide rapidity region of $-2.7 < y < 2.7$. These Si detectors will provide a multiplicity trigger. Also, at forward backward angles a magnetic field perpendicular to the beam direction will be created by a piston-type magnet and, there, muons will be measured. The number of total charged particles measured by PHENIX is only 300–500 which is by one order of magnitude smaller than the number of charged particle tracks measured by STAR. However, PHENIX has a strong capability of particle identification not only for electrons and muons but also for photons and hadrons.

Both STAR and PHENIX detectors together with their kinematical coverage are shown in Fig. 18. Two detectors will start to take data in 1999.

*3.4. LHC Experiment*

At LHC (Large Hadron Collider) at CERN a heavy-ion acceleration is also planned. The beam energy for Pb + Pb collisions is 3.8 A•TeV in a collider mode. There, a new experiment, called ALICE, is approved. The construction of this detector will start soon. The ALICE has TPC (like STAR at RHIC) and also it has forward and backward muon arms (like PHENIX at RHIC). Since the energy density achieved at LHC is much higher than that at RHIC, it is very intriguing to see how each signature of QGP exhibits at the LHC energy region. The data taking will start, most likely, in 2005.

## 4. Summary

At the beginning of this talk I described two fundamental questions which have been raised in nuclear and particle physics: the identity of hadrons in nuclear matter and the quark confinement. In order to solve these questions, the study of nuclear matter at extreme densities and temperatures, in particular, the latter, is important. It is likely that a new phase called the quark-gluon plasma is formed at high temperatures. In the next century, experimental studies on the properties of the quark-gluon plasma will be the key element in the field of relativistic heavy-ion physics. These studies will provide the most exciting physics in both nuclear and particle physics.

At the same time, several interesting data have been generated from the BNL-AGS and the CERN-SPS, such as the study of hadron properties at finite nuclear-matter density. Systematic studies of properties of nuclear matter at high density is also an important task in the next century and they should also be pursued.

Finally, the author would like to express his sincere appreciation to the organizers of this Symposium for their kind invitation. Also, he would express his thanks to H. Specht and J. Stachel for providing me data and figures.

# References

1. F. R. Brown, et al., Phys. Rev. Lett. **65**, 2491 (1990); S. Chandrasekhara, Nucl. Phys. **B42**, 475 (1995).
2. K. Kanaya, Nucl. Phys. **B47**, 144 (1996), and references therein.
3. W. Weise, Nucl. Phys. **A443**, 59c (1993), and references therein.
4. Y. Nambu and G. Jona-Lasino, Phys. Rev. **122**, 345 (1961).
5. K. Shigaki, Ph.D Thesis, University of Tokyo (1995).
6. P. J. Siemens and J. O. Rasmussen, Phys. Rev. Lett. **42**, 844 (1979).
7. J. Stachel, Nucl. Phys. **A610**, 509c (1996).
8. L. D. Landau, Izv. Acad. Nauk SSSR Ser. Fiz **17**, 51 (1953); A. S. Goldhaber, Nature **275**, 114 (1978).
9. R. Mattiello, H. Sorge, H. Stoecker and W. Greiner, Phys. Rev. Lett. **63**, 1459 (1989); Y. Pang, T. J. Schlagel and S. H. Kahana, Phys. Rev. Lett. **68**, 2743 (1992); Y. Pang, CCAST Symposium/Workshop Proceedings **10**, 451 (1993), Gordon & Breach Scientific Publishers, edited by Y. Pang, J. W. Qui and Z. M. Qiu.
10. Y. Akiba, et al. (E802 Collaboration), Phys. Rev. Lett. **76**, 2021 (1996.
11. T. Ullrich, et al. (NA45 Collaboration), Nucl. Phys. **A610**, 317c (1996); A. Dress, Nucl. Phys. **A610**, 536c (1996).
12. T. Matsui and H. Satz, Phys. Lett. **B178**, 416 (1986).
13. M. Gonin, et al. (NA50 Collaboration), Nucl. Phys. **A610**, 405c (1996).
14. C. Gerschel and J. Hufner, Z. Phys. **C56**, 71 (1992).
15. J. D. Bjorken, Phys. Rev. **D27**, 140 (1983).
16. L. Van Hove, Phys. Lett. **B118**, 138 (1982).
17. W. V. Jones, Y. Takahashi, B. Wosiek, and O. Miyamura, Ann. Rev. Nucl. Part. Sci. **37**, 71 (1987)..
18. K. Rajagopal and F. Wilczek, Nucl. Phys. **B399**, 395 (1993), F. Wilczek, Nucl. Phys. **A566**, 123c (1994).
19. See, for example, P. V. Ruuskanen, Nucl. Phys. **A544**, 169c (1992).

# PRESENT AND FUTURE OF HIGH ENERGY PHYSICS WITH HADRON COLLIDERS

K. KONDO

*Institute of Physics, University of Tsukuba,*
*Ibaraki 305, Japan*

Present status and future plans of high energy physics with hadron colliders are reviewed. Observation of top quark, recent results on bottom quark physics and QCD phenomena are discussed. Current mass limits and future plans of search for Higgs particle, supersymmetric and other new particles are described.

## 1  Introduction

Experimental high energy physics today is largely benefitted by two types of accelerators, $e^+e^-$ and $\bar{p}p$ colliders. Accelerator techniques are well advanced and sophisticated respectively, providing valuable information on particle physics. In this report, subjects are limited mainly to physics with hadron colliders, simply because the author is familiar with them.

Usefulness of hadron colliders in exploring high energy frontiers has been demonstrated by observations of $W$ and $Z$ bosons[1][2] and top($t$) quark. Current activities at Tevatron have provided detailed information on electroweak physics with $W$ and $Z$ bosons, production and decay of charm($c$) and bottom($b$) quarks, test of Quantum Chromodynamics(QCD) with jet phenomena and top quark physics. Preparation of a next generation $pp$ collider, Large Hadron Collider(LHC) at CERN, is on its way aiming at searches for new particles.

## 2  Observation of Top Quark

Top quark is the 6-th quark in the Standard Model[3]. The first evidence for the $t\bar{t}$ production was reported by CDF collaboration in 1994[4], and confirmations were made by CDF and D0 groups in 1995[5][6]. Up to present, data of approximately 100 pb$^{-1}$ were collected in each experiment. We discuss into some details top quark physics as a current topical subject.

### 2.1  Top quark signatures and observed numbers of event

Signatures of the $t\bar{t}$ production depend primarily on the decay modes of the $W$'s: (a) "dilepton" channel, where both $W$'s decay as $W \to l\nu$ and the final state is $l^+\nu l^-\nu b\bar{b}$, (b) "Lepton + jet" channel, where one $W$ decays as $W \rightarrow l\nu$ and the final state is $l^{+/-}\nu\ q\bar{q}'\ b\bar{b}$, (c) "multi-jet" channel, where both $W$'s decay as $W \rightarrow q\bar{q}'$ and the final state is $q_1\bar{q_1}'\ q_2\bar{q_2}'\ b\bar{b}$. Here $l$ is either $e$ or $\mu$, and the branching ratios for three channels are (a) 5%, (b) 30%, and (c) 44%, respectively.

(a) Dilepton channel    Signatures for this channel are two isolated high $P_T$ leptons($e$, $\mu$, $\tau$), missing transverse energy $\not{E}_T$ and two or more jets. Dominant background are $WW$ production, $Z \rightarrow \tau\tau$, Drell-Yan process and fake leptons. This channel has a good signal-to-background ratio but low statistics.

(b) Lepton($l$)+Jets channel    The $t\bar{t}$ signatures for this channel include one isolated high $P_T$ lepton ($e$ or $\mu$), missing $E_T$ and 4 or more jets, two of which are from $b$−quarks. Dominant backgrounds for this channel are $\bar{p}p \rightarrow W$+jets and QCD backgrounds one particle in which is faking a lepton.

Reductions of backgrounds are made in different ways: (i) Cuts on event shape, i.e. aplanarity, $H_T = \sum E_T$(jets) and $t\bar{t}$ vs. background likelihood formed on the basis of jet $E_T$'s, (ii) tagging of $b$−quarks by use of semileptonic decay, $b \rightarrow \mu X$ (20%) and $b \rightarrow eX$ (20%), and (iii) tagging of $b$−quarks with displaced vertex (CDF).

(c) Multijet channel    The signature of $t\bar{t}$ production in the multijet channel is an existence of 6 or more jets, two of which are from $b$−quarks. Dominant backgrounds are QCD multijet production. Signal-to-background ratio in this channel is approximately 1/30 before additional topological/kinematical requirements or $b$−tagging.

In total, we have approximately 13 dilepton, 70 $l$+jets and 60 multijet events in the CDF and D0 collaborations.

## 2.2    Production and decay properties

Cross section    The measurement of $t\bar{t}$ production cross section has been made in each decay channel individually, and compared with one another and to theoretical prediction (see Fig. 1).

CDF measurement of $V_{tb}$    Unitarity within the three-generation Standard Model implies $V_{tb} \approx 1.0$. CDF has analyzed the $l$+jets and dilepton samples to measure the ratio of events with 0, 1 and 2 $b$−tags and use it to extract $b \equiv$ Br($t \rightarrow Wb$)/Br($t \rightarrow WX$).

Results of a maximum likelihood combining all information yields $b = 0.94 \pm 0.27$(stat.) $\pm 0.13$(syst.), or $b > 0.34$ at 95% C.L. Assuming three-generation unitarity, one obtains $|V_{tb}| = 0.97 \pm 0.15 \pm 0.07$, and $|V_{tb}| > 0.58$ at 95% C.L.

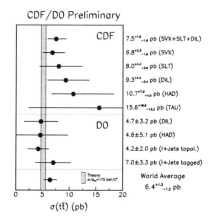

CDF/D0 Preliminary

7.5$^{+1.9}_{-1.8}$ pb (SVX+SLT+DIL)

6.8$^{+3.3}_{-1.8}$ pb (SVX)

8.0$^{+4.4}_{-3.6}$ pb (SLT)

9.3$^{+4.4}_{-3.4}$ pb (DIL)

10.7$^{+7.4}_{-4.0}$ pb (HAD)

15.6$^{+10.4}_{-13.2}$ pb (TAU)

4.7±3.2 pb (DIL)

4.6±5.1 pb (HAD)

4.2±2.0 pb (l+Jets topol.)

7.0±3.3 pb (l+Jets tagged)

World Average
6.4$^{+1.3}_{-1.2}$ pb

Theory
at M$_{t}$=175 GeV/c$^2$

σ(tt̄) (pb)

Figure 1: Summary of the cross section measurements and the theoretical value.

Search for rare decays
A search has been made for $t\bar{t}$ events with one top decaying in the standard mode $(t \rightarrow Wb)$ and the other decaying to a rare mode: $t \rightarrow \gamma c$, $t \rightarrow \gamma u$(B.R.$\approx 10^{-10}$) and $t \rightarrow WZb$. Theoretically one expects no event and limit will not be a strong test of the Standard Model. A preliminary search for $t \rightarrow \gamma q$ by CDF yields B.F.$[(t \rightarrow \gamma c)+(t \rightarrow \gamma u)] < 2.9\%$(95%C.L.).

## 2.3  Mass measurement

The mass measurement of top quark is important since it determines one of the Standard Model parameters. The mass can be determined in all decay modes. Most favorable channel for the mass measurement, however, is the $l$+jets channel, because of a reasonably good signal-to-background ratio and kinematical constraints.

D0 $l$+jets mass determination  Event selections are made by requiring 4 or more jets in the soft muon $b$−tagged sample or $|\eta_W| < 2.0$ in the non-tagged sample. A total of 8 tagged events and 85 non-tagged events were obtained.

To separate top from the background, they define 4 kinematical variables using Signal Monte Carlo events. Signal events are shown to be accurately selected with these variables using the control sample. The 4 variables are combined into one discriminant $D$ which separate signal from background. It is also confirmed that $D$ is almost completely independent of the top mass. A preliminary result of D0 mass fit is shown in Fig. 2. The mass of top quark is 169±8(stat.) GeV/c$^2$ with a fit $\chi^2$ of 21.7/22.

CDF $l$+jets mass measurement  CDF attempts to fully reconstruct the top quarks from their decay products in a probabilistic way. For this purpose one assumes that the 4 leading−$E_T$ jets correspond to $b$, $\bar{b}$ and $q\bar{q}'$ from

142

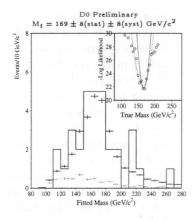

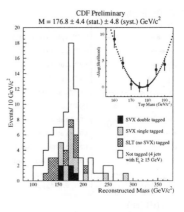

Figure 2: Fitted mass by D0 likelihood method.

Figure 3: Reconstructed mass by the CDF optimized method.

$W$. Unmeasured $z$−component of the neutrino momentum is determined from a constraint $M_{l+\nu} = M_W$ with two-fold ambiguities. There are 24 combinations without $b$−tagging, 12 with one $b$−tagged, and 4 with 2 $b$−tagged events. For full reconstruction, one applies two constraints; $M_{jj} = M_W$ and $M_{top} = M_{antitop}$. By forming a $\chi^2$, the best jet assignments and $P_z(\nu)$ solution are selected.

CDF has recently improved the mass determination strategy by dividing samples according to their qualities. In this optimized method, they treat 1 SVX tagged events, 2 SVX tagged events, SLT- but not SVX-tagged events and non-tagged events, separately. Each sub-sample is fitted to Monte Carlo events of top signal and background by varying $M_{top}$. Then a likelihood against $M_{top}$ is extracted in each sub-sample. Since they are statistically independent, product of the sub-sample likelihoods is used to get one result. Figure 3 shows preliminary results by the optimized method.

Systematic uncertainties    Systematics on the $l$+jets samples of CDF and D0 are listed in Table 1. The dominant uncertainty is in the energy scale of jet where how much fractions of soft gluons from $t\bar{t}$ and QCD are included in a jet is not well understood. Also hard gluons which are radiated from the top decay products but identified as another jet bring in different sort of uncertainty.

Summary of mass measurement    Figure 4 shows the top mass measured by various methods. For mass value at present, we use only $l$+jets sample. If we combine the CDF result with the D0 result, the mass is $M_{top} = 175 \pm 6$ GeV/$c^2$, folding the statistical and systematic errors.

Table 1 Mass systematics on the Lepton+Jets sample.

CDF/DO Preliminary

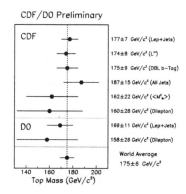

Figure 4: Top quark mass measured by various methods.

### CDF Preliminary

| Items | $(GeV/c^2)$ |
|---|---|
| Soft gluon + Jet $E_T$ Scale | 3.6 |
| Different generators | 1.4 |
| Hard gluon effects | 2.2 |
| Kinematic/Likelihood Fitting | 1.5 |
| $b$−tagging bias | 0.4 |
| Background spectrum | 0.7 |
| Monte carlo statistics | 0.8 |
| Total | 4.8 |

### DO Preliminary

| Items | $(GeV/c^2)$ |
|---|---|
| Jet energy correction | 7.3 |
| Monte carlo model | 3.3 |
| Fitting method | 2.0 |
| Total | 8 |

## 2.4  Summary on top quark observation

- Top has been observed and $\sigma_{t\bar{t}}$ has been measured in many decay modes; $t\bar{t} \rightarrow W+\text{Jets}+X$, $t\bar{t} \rightarrow W + b+\text{Jets}+X$, $t\bar{t} \rightarrow l^+l^-+\text{Jets}+X$, $t\bar{t} \rightarrow l + \tau+\text{Jets}+X$, $t\bar{t} \rightarrow 6\text{Jets}+X$.

- World average $t\bar{t}$ cross section at 175 GeV/c$^2$ is $\sigma_{t\bar{t}}= 6.4\ ^{+1.3}_{-1.2}$ pb. QCD predictions range from 4.7 to 5.6 pb.

- New top mass measurements emphasize optimal use of information. Current values determined by $l$+jets events are 169±11 GeV/c$^2$ (D0), and 176.8±6.5 GeV/c$^2$ (CDF). Averaging the two, we obtain $M_{top}$ =175±6 GeV/c$^2$.

Nothing observed in top production or decay is strongly inconsistent with the Standard Model.

In passing, the author would like to emphasize the importance of experimental techniques developed for the observation of top quark. Experimentalists identified jets with daughter quarks by aids of $b$−tagging techniques or by applying kinematical constraints, and made full reconstructions of events. They are beginning to deal with jets as daughter quarks and reconstruct events at the quark-lepton level as theorists describe them.

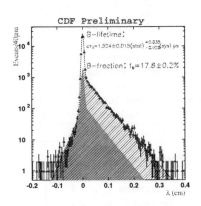

**Figure 5:** Pseudo-proper time distribution in $B^0$ decay

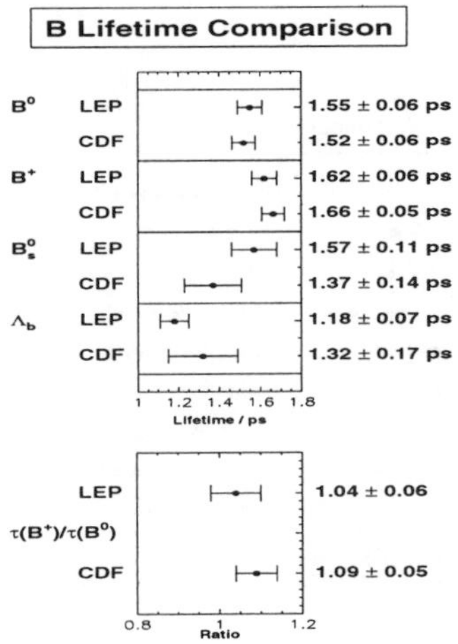

**Figure 6:** Summary of life time measurements of $B$-meson.

# 3   $B$–quark Physics

Bottom quark physics is one of the popular subjects in the present and near future high energy physics. Constructions of $B$–factories of the $e^+e^-$ machines are in progress. A major goal is observation of $CP$–violation in the b-quark. Here we review what have been observed on $b$–quark with an existing collider and discuss the prospect of $b$–physics with hadron colliders in the near future[7]. An advantage of pursuing $b$–physics with hadron colliders is in a high production rate, and it is important to establish the $b$ identification from a huge background in the hadron collider environment.

<u>Observation of $B$ mesons</u>   Observations of known B-mesons $B^{0,\pm}$ which consists of $b$ and $u$ or $d$ quarks have been established, and the precise measurements of their masses and lifetimes have been made. The key experimental techniques are to find $b$'s in either lepton($e$ or $\mu$) inclusive events or by full reconstruction of $B$ mesons from their daughter particles, e.g. $B^+ \rightarrow J/\psi K^+$ or $B^0 \rightarrow J/\psi K^{*0}$. Figure 6 shows the results of $b$ lifetime measurements by LEP and Tevatron experiments. The lifetimes are measured from the flight distances of $B$ mesons from the primary vertex(see Fig. 5).

New hadrons including $b$–quark, $\Lambda_b$ and $B_s$ have been observed at Tevatron. Searches for $B_c$ are also in progress.

<u>$B^0\bar{B}^0$ mixing</u>   $CP$ eigenstates of neutral B-meson are thought to be superpositions of states at production, $B^0$ and $\bar{B}^0$. As a consequence, $B^0 - \bar{B}^0$

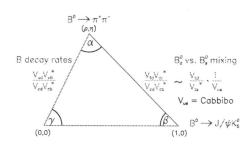

Figure 7: Unitarity triangle of CKM matrix elements.

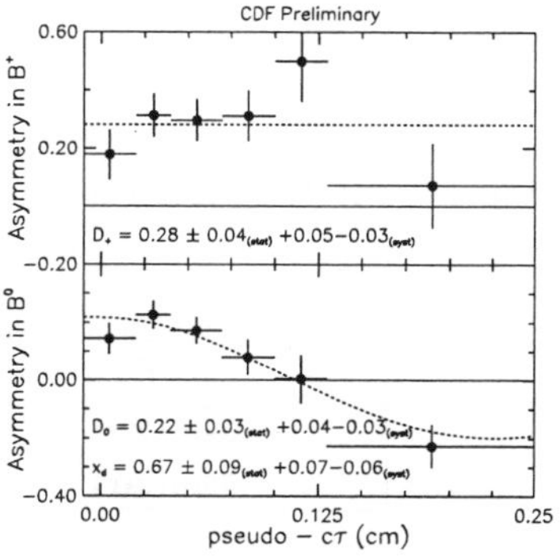

Figure 8: $B^0 - \bar{B}^0$ mixing compared to $B^+$ decay.

conversion occurs during the decay of neutral $B$−mesons. Observations of the time integrated conversion effect have been made by identifying $B^0/\bar{B}^0$ with $l^-/l^+$ in the semileptonic decay modes of neutral $B$−mesons.

*CP violation*   $CP$ violation in the $b$−quark sector with the three-generation Standard Model[8] is represented by the unitarity triangle in the complex plane shown in Fig. 7. Oblique sides are given by the ratios of products of Cabbibo-Kobayashi-Maskawa(CKM) matrix elements. Thus angles correspond to phase differences of the matrix elements. Experimentally angles $\alpha$ and $\beta$ and the length of right oblique side are planned to be measured.

$CP$ violation in $b$−quark interaction can be observed from difference in the decay rates for $B^0$ and $\bar{B}^0$. To observe the $CP$ violation in the $b$−quark system identification of $B^0$ vs. $\bar{B}^0$ is crucial.

Tagging of $b$ vs. $\bar{b}$ can be made by knowing charge-signs of particles involved: (a) jet charge, *i.e.* the momentum-weighted sum of the charge of tracks recoiling against a $B$ meson, (2) charge of a lepton from the second $b$, (3) charge correlations between the $B$ meson and charged tracks in its vicinity.

Figure 8 shows an example of $B^+$ and $B^0$ proper time evolution. Here, flavors are tagged using a same side pion charge, and by reconstructing the $D^0$ or $D^{*+}$ meson from the semileptonic decay on the opposite side. A clear mixing signal is present for the neutral $B$ while no mixing is observed for charged $B$

meson. This example indicates the feasibility of flavor tagging for observation of $CP$ violation.

Table 2 Expected accuracies in the measurements of $CP$ violating angles.

|  | CDF($\bar{p}p$) | BaBar($e^+e^-$) |
|---|---|---|
| Integrated Luminosity | 2 fb$^{-1}$ | 30 fb$^{-1}$ |
| $\epsilon D^2$ | $2.7 - 2.8\%$ | $13 - 34\%$ |
| $\delta\sin(2\beta)$ | 0.08- 0.16 | 0.059 |
| $\delta\sin(2\alpha)$ | 0.10-0.14 | 0.085 |

A dilution factor $D$ due to miss-tagging and the detection efficiency $\epsilon$ determines the accuracy in the measurements of $CP$ violating effects. Table 2 shows a comparison of prospects of the measurements on $\sin(2\beta)$ and $\sin(2\alpha)$ in typical planned experiments at $\bar{p}p$ and $e^+e^-$ colliders.

# 4 QCD Phenomena

Jets as QCD phenomena are most copiously produced in hadron colliders. Experimental measurements are now precise enough to make serious tests of the theory[7].

Jet inclusive cross section    Recently the inclusive jet cross section has been measured at Tevatron and agreement between theory[9] and experiment extends over about $10^{10}$ range of the cross section(see Fig. 9). At high $E_T$ region, however, the observed cross section shows an excess over the QCD cross section as shown in Fig. 10. The result might suggest compositeness, but explanation within the Standard Model, e.g. by modification of the parton distribution function, cannot be rejected at the present stage.

Running coupling constant $\alpha_s$    In the Standard Model of present particle physics, coupling constants of the electroweak and strong interactions $\alpha$ and $\alpha_s$ are functions of the energy scale representing the process.

From the jet inclusive cross section measured at CDF, value of $\alpha_s$ was extracted by theorists[14] as a function of $E_T$ of jet (see Fig. 11. The result suggests a rise of $\alpha_s$ near $E_T = 300$ GeV. Updating efforts by experimenters are in progress with more data recently collected.

Heavy quark production    Production cross sections of heavy quarks $c$ and $b$ have been measured at Tevatron. Figure 12 shows the $b$−quark inclusive cross section , $\bar{p}p \to bX$. The identification of $b$ was made in various decay channels;

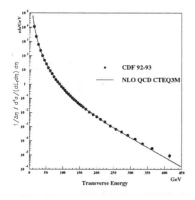

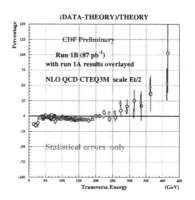

Figure 9: Jet inclusive cross section.

Figure 10: High $E_T$ anomaly in jet inclusive cross section

electron and muon inclusive channels $e^- X$ and $\mu^- X$, and channels including charm quark, $e^- D^0$, $\psi X$ and $\psi' X$. Results are compared with next-to-leading-order(NLO) QCD calculations. Apparently the measured cross section shows excess over the theoretical calculations[9].

Another example of a cross section where the measured value is more than the calculated is in the production of $\psi'$[10].

We have emphasized measurements which do not show good agreements with calculations. A personal prejudice of the author is that the window for beyond-the-Standard-Model physics could be opened in the strong interaction sector.

## 5 New Particle Searches

The Standard Model has been so successful that no contradicting experimental evidence has been reported for past three decades. Nevertheless, it could not make theorists happy since its birth. Supersymmetry theory[11], Technicolor model[12] and quark-lepton compositeness[13] are among challenging theoretical scenarios beyond the Standard Model.

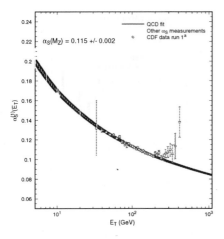

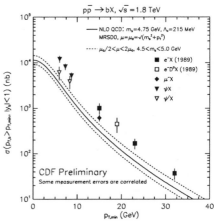

Figure 11: Running strong coupling constant $\alpha_s$.

Figure 12: $b\bar{b}$ production cross section

## 5.1 Higgs particle

We start with unobserved particle within the Standard Model. The gauge symmetry and/or chiral symmetry requires mass of particles to be zero. Higgs particle $H^0$ is an important element of the Standard Model proposed as the dynamical origin of mass of particles. The theory assumes vacuum to be a condensate of Higgs particles. This reminds us of history of filling vacuum with ether or negative energy electrons. Whether true or not, experimentalists should search for this particle. Actually, in many cases a justification of future gigantic accelerators is in Higgs searches.

For a low or intermediate mass($m_H < 170$ GeV/c$^2$), Higgs particle is expected to be produced associated with weak bosons; $e^+e^- \rightarrow ZH$ at LEP, and $q\bar{q}' \rightarrow WH$ at Tevatron[7]. Decay channel of Higgs particle are $H \rightarrow b\bar{b}$(80%) or $\tau^+\tau^-$(7%). When Higgs mass is higher than 170 GeV/c$^2$, production by a gluon fusion $gg \rightarrow H$ and decay of $H \rightarrow ZZ^{(*)} \rightarrow l^+l^-l^+l^-$ is thought to be a favorable channel. We quote some details of LHC plans[15] of Higgs searches.

Intermediate mass Higgs at LHC  For the direct production mode, reliable signatures are $H \rightarrow \gamma\gamma$ and $H \rightarrow 4$ leptons. For associated production modes, low rate but easier mode is $W + H, t\bar{t}; H \rightarrow \gamma\gamma$, and rather difficult mode is $W + H, t\bar{t}; H \rightarrow b\bar{b}$. Table 3(a) summarizes expected numbers of yield with an integrated luminosity of $10^5$ pb$^{-1}$(one year of running with an instantaneous luminosity of $10^{34}$sec$^{-1}$cm$^{-2}$).

Heavy mass Higgs at LHC    For heavy mass of Higgs, $H \to WW \to l\nu jj$ and $H \to ZZ \to lljj$ are favorable decay channels. The estimated numbers of signal and background events are listed in Table 3(b), assuming $m_H = 1$ TeV/c$^2$ and the integrated luminosity of $10^5$ pb$^{-1}$ and lepton detection efficiency of 90%.

Table 3(a) Expected numbers of events in search for intermediate mass Higgs particles at LHC(Integrated luminosity = $10^5$ pb$^{-1}$).

| $m_H$ | $\gamma\gamma$ | $4l$ | $WH$ $+t\bar{t}H$ |
|---|---|---|---|
| 80 | 2.2 | | 3.9 |
| 110 | 5.1 | | |
| 150 | 4.2 | 21.7 | |

Table 3(b) Expected numbers of events in search for high mass Higgs particles at LHC (Integrated luminosity = $10^5$ pb$^{-1}$).

| Process | Jet veto | Single tag | Double tag |
|---|---|---|---|
| $H \to WW$ | 251 | 179 | 57 |
| $t\bar{t}$ | 560 | 110 | 5 |
| $W$+jets | 3820 | 580 | 12 |
| Pile-up | | 160 | 2 |
| $S/\sqrt{B}$ | 3.8 | 6.8 | 13.8 |

## 5.2 $W'$ and $Z'$

Heavy neutral gauge bosons $Z'$ occur in any extension of the Standard Model that contains an extra U(1) sector. For example, in one model with $E_6$ as the grand unified gauge group[16] there exist $Z_\psi$ and $Z_\chi$ from the symmetry breaking $E_6 \to SO(10) \times U(1)_\psi \to SU(5) \times U(1)_\chi \times U(1)_\psi$. In superstring inspired $E_6$ models there exists a $Z_\eta = \sqrt{3/8}\, Z_\chi + \sqrt{5/8} Z_\psi$.

Experimental searches for $Z'$ have been made with $Z' \to ll$ where $l = e$ or $\mu$. Assuming the Standard Model coupling for $Z' \to ll$ decay, the current mass limit of $M_{Z'} > 690$ GeV/c$^2$ has been obtained[7].

Heavy W bosons, $W'$, occur, for example, in the left-right symmetric model of electroweak interactions $SU(2)_R \times SU(2)_L \times U(1)_Y$[17]. With the search in the $W' \to l(e$ or $\mu)\nu$ mode the mass limit of $M_{W'} > 652$ GeV/c$^2$ has been obtained, assuming Standard Model coupling[7].

Undoubtedly, the mass ranges of searches can be extended by several times at LHC.

## 5.3 Supersymmetric particles

In the Standard Model, "matter" particles, quarks and leptons, are fermions, and "force" particles, photon, $W$, $Z$ and gluon, are bosons. Theory of Supersymmetry(SUSY) unifies the matter and force, and claims existence of super-

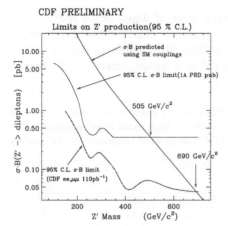

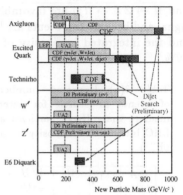

Figure 13: $Z'$ search limit.

Figure 14: Excluded mass regions for new particles.

partners of known particles. Supersymmetric partners and Higgs bosons are listed in Table 4. Current mass limit for these particles are shown in Table 5.

Among Supersymmetric Grand Unified Theory(SUSY-GUT), Minimal Supersymmetric Standard Model(MSSM)[18] gives mass spectrum of SUSY particles as a function of 5 parameters, $m_0$, $m_{1/2}$, $A_0$, $\tan\beta$ and $\mu$. Evaluating these parameters from existing data by use of renormalization technique, many models along MSSM predict rather low-lying superpartners, lighter than a few hundred GeV/c$^2$, which can be reached with upgraded Tevatron.

Table 4 Supersymmetric partners and Higgs bosons

| Particle Name | Spin | Physical States |
|---|---|---|
| squarks | 0 | $\tilde{d}_L, \tilde{u}_L, \tilde{s}_L, \tilde{c}_L, \tilde{b}_1, \tilde{t}_1, \tilde{d}_R, \tilde{u}_R, \tilde{s}_R, \tilde{c}_R, \tilde{b}_2, \tilde{t}_2$ |
| slepton | 0 | $\tilde{e}_L, \tilde{\nu}_{eL}, \tilde{\mu}_L, \tilde{\nu}_{\mu L}, \tilde{\tau}_1, \tilde{\nu}_{\tau L}, \tilde{e}_R, \tilde{\mu}_R, \tilde{\tau}_2$ |
| charginos | 1/2 | $\tilde{\chi}_1^\pm, \tilde{\chi}_1^\pm$ |
| neutralinos | 1/2 | $\tilde{\chi}_1^0, \tilde{\chi}_2^0, \tilde{\chi}_3^0, \tilde{\chi}_4^0$ |
| gluino | 1/2 | $\tilde{g}$ |
| Higgs bosons | 0 | $h, H, A, H^\pm$ |

Since SUSY particles has a new quantum number $R$, they are expected to be produced in pairs. The key signature is the missing transverse energy carried away by the lightest supersymmetric particle $\tilde{\chi}_1^0$. By transferring $R$ quantum number to $\tilde{\chi}_1^0$ the superpartners can decay as its known partners.

Table 6 shows key decay modes for SUSY particles and experimental signatures to detect them. The difficulty in SUSY search is their low production cross sections. The full event reconstruction technique developed for observations of $t\bar{t}$ events will be useful in the SUSY searches.

Table 5 Current mass limit on supersymmetric partners

| Sparticle | Mass Limit | Comments |
|---|---|---|
| $\tilde{g}$ | 160 GeV | CDF & D0 |
| $\tilde{q}$ | 220 GeV | CDF & D0 |
| $\tilde{t}_1$ | 45 GeV | LEP |
| $\tilde{\chi}_1^{\pm}$ | 47 GeV | LEP |
| $\tilde{\chi}_1^0$ | 20 GeV | LEP |
| $\tilde{l}$ | 45 GeV | LEP |
| $\tilde{\nu}$ | 43 GeV | LEP |
| $h$ | 60 GeV | LEP |

Table 6 Decay modes and signatures of supersymmetric partners

| Production | Key Decay Mode | Signature |
|---|---|---|
| $\tilde{g}\tilde{g}, \tilde{g}\tilde{q}, \tilde{q}\tilde{q}$ | $e.g., \tilde{g} \to qq\tilde{\chi}_1^0, \tilde{q} \to qqq\tilde{\chi}_1^0 (M_{\tilde{q}} > (M_{\tilde{g}}$ | $\not{E}_T$ + multijets |
| $\tilde{\chi}_1^{\pm}\tilde{\chi}_2^0$ | $\tilde{\chi}_1^{\pm} \to \tilde{\chi}_1^0 l\nu, \tilde{\chi}_2^0\tilde{\chi}_1^0 ll$ | Trilepton + $\not{E}_T$ |
| | $\tilde{\chi}_1^{\pm} \to \tilde{\chi}_1^0 qq, \tilde{\chi}_2^0\tilde{\chi}_1^0 ll$ | Dilepton + $\not{E}_T$ + jets |
| $\tilde{t}_1\tilde{t}_1$ | $\tilde{t}_1 \to \tilde{\chi}_1^{\pm} b, \tilde{\chi}_1^{\pm} \to \tilde{\chi}_1^0 l^{\pm}\nu, \tilde{\chi}_1^{\mp} \to \tilde{\chi}_1^0 qq$ | Single lepton + $\not{E}_T$ +b's |
| | $\tilde{t}_1 \to \tilde{\chi}_1^{\pm} b, \tilde{\chi}_1^{\pm} \to \tilde{\chi}_1^0 l^{\pm}\nu, \tilde{\chi}_1^{\mp} \to \tilde{\chi}_1^0 l^{\mp}\nu$ | Dilepton + $\not{E}_T$ + b's |
| $W/Z + h$ | $h \to bb, \tau\bar{\tau}$ | 2 b's or 2 $\tau$'s |

## 5.4 Technicolor scenario

Recently, there have been much efforts on Technicolor theories [19][20]. Topcolor model[21] was proposed to overcome major shortcomings of top-condensate models of Higgs particle[22]. An attractive feature of the Topcolor model to experimentalists is that it predicts new strong gauge bosons, *i.e.* massive gluon, or mesons with rather low masses, $m \approx 300$ GeV/$c^2$. Thus, they can be serious targets with hadron colliders in the near future[7].

# 6   Summary

We have reviewed present status of hadron collider physics and discussed planned experiments in the near future.

- Top quark has been observed at Tevatron. There is no strong evidence against the Standard Model at present. Precise measurements at high luminosity Tevatron and LHC will continue further studies on top to see if there is any indication beyond the Standard Model.

- $B$-quark identifications and precise measurements of mass and lifetime in the hadron collider environments have been demonstrated by recent CDF experiments. $CP$ violation is a major goal of $e^+e^-$ $B$-factories and hadron colliders. Hadron collider experiments will provide information complimentary and competitive to $B$-factories on this subject.

- Experimental tests on QCD with high $E_T$ jets are precise enough to confront with the theory. There are some preliminary results showing quantitative disagreements between theory and experiment. Further studies on high $E_T$ jets with future hadron colliders could lead us to some new physics.

- Searches of Higgs particle is another goal of testing the Standard Model. For low mass($< 150$ GeV/c$^2$) Higgs, Tevatron as well as LHC has capabilities to detect. For higher mass Higgs up to 1 TeV/c$^2$, LHC has a great chance to discover.

- Searches for new particles predicted by theory; Supersymmetric partners, new electroweak gauge bosons, Topcolor or Technicolor and new types of quarks or hadrons associated with them, will be pursued with Tevatron and LHC. Compositeness of quarks and leptons is another item to be investigated with these machine.

We wish and anticipate to uncover some new physics beyond the Standard Model in the next few decades.

# References

[1] UA1 Collaboration, Phys. Lett. **166B**, 484 (1986); **185B**, 233 (1987).

[2] UA2 Collaboration, Z. Phys. **C30**, 1 (1986); Phys. Lett. **186B**, 440, 452 (1987).

[3] S. Glashow, Nucl. Phys. **22**, 579 (1961); S. Weinberg, Phys. Rev. Lett. **19**, 1264 (1967); A. Salam, *Weak and Electromagnetic Interactions in Elementary Particle Theory*, W. Svartholm, ed., (Almquist and Wiksell, Stockholm 1968).

[4] CDF Collaboration, Phys. Rev. **D50**, 2966 (1994).

[5] CDF Collaboration, Phys. Rev. Lett. **74**, 2626 (1995).

[6] D0 Collaboration, Phys. Rev. Lett. **74**, 2632 (1995).

[7] CDF COllaboration, *The CDF II Detector*, November 1996 (Fermilab).

[8] M. Kobayashi and T. Maskawa, Prog. Theor. Phys. **49**, 652 (1973). For a review of $CP$ violation in $B$ decays, Y. Nir and H. R. Quinn, Ann. Rev. Nucl. and Part. Sci. **42**, 211 (1992).

[9] S. Ellis, Z. Kunst and D. Soper, Phys. Rev. **D40** 2188 (1980).

[10] M. Schmidt, *4th Topical Conference on Flavor Physics*, KEK, Japan, Oct 29-31, 1996.

[11] D.V. Volkov and V.P. Akulov, Phys. Lett. **46B**, 109 (1973); J. Wess and B. Zumino, Nucl. Phys. **B70**, 39 (1974).

[12] S. Weinberg, Phys. Rev. **D13**, 974 (1976); Phys. Rev. **D19**, 1277 (1979). L. Susskind, Phys. Rev. **D20**, 2619 (1979).

[13] J.C. Pati and A. Salam, Phys. Rev. **D10**, 275 (1974); H.Terazawa, Y. Chikashige and K. Akama, Phys. Rev. **D15**, 480 (1977); H. Terazawa, Phys. Rev. **D22**, 184 (1980).

[14] W.T. Giele and E.W.N. Glover, Moriond 95.

[15] ATLAS Collaboration Design Report; CMS Collaboration Design Report, CERN.

[16] See F. del Aguila, M. Quiros and F. Zwirner, Nucl. Phys. **B287**, 457 (1987) and references therein.

[17] For a review and original references see R.N. Mahapatra, *Unification and Supersymmetry*, (Springer, New York, 1986).

[18] S. Dimopoulos and H. Georgi, Nucl. Phys. **B193**, 150 (1981): N. Sakai, Z. Phys. **C11**, 153 (1981). For a review see for example H.E. Haber and G.L. Kane, Phys. Reports **117**, 75 (1985).

154

[19] K. Lane and M.V. Ramana, Phys. Rev. **D44**, 2678 (1991); E. Eichten and K. Lane, Phys. Lett. **B327**, 129 (1994).

[20] T. Appelquist and G. Triantaphyllou, Phys. Rev. Lett. **69**, 2750 (1992); T. Appelquist and J. Terning, Phys. Rev. **D50**, 2116 (1994).

[21] C.T. Hill, Phys. Lett. **266B**, 419 (1991); C.T. Hill and S. Parke, Phys. Rev. **D49**, 4454 (1994). For a review see for example K. Lane, BUHEP-96-8(1996).

[22] Y. Nambu, in *New Theories in Physics*, Proceedings of the XI International Symposium on Elementary Particle Physics, Kazimierz, Poland, 1989);V.A. Miransky, M. Tanabashi and K. Yamawaki, Phys. Lett. **B221**, 177 (1989); W.A. Bardeen, C.T. Hill and M. Linder, Phys. Rev. **D41**, 1647 (1990).

# COSMOLOGICAL PARAMETERS AND EVOLUTION OF GALAXIES: AN OBSERVATIONAL PERSPECTIVE

Sadanori OKAMURA

*Department of Astronomy and Research Center for the Early Universe,*
*School of Science, University of Tokyo*
*Bunkyo-ku, Tokyo 113, Japan*

According to the Friedmann models without the cosmological constant, structure and time evolution of the Universe are characterized by the two parameters, the Hubble constant $H_0$ and the density parameter $\Omega_0$. Recent observational determinations of the Hubble constant is reviewed and $H_0 \sim 73 \pm 10$ kms$^{-1}$Mpc$^{-1}$ is shown to be the present best estimate of the local ($\lesssim 100$ Mpc) value. The determination of the density parameter is shown to be coupled with the evolutionary history of galaxies, which is not well understood yet. The density parameter is not constrained by the observation very well at present and $0.1 < \Omega_0 \leq 1.0$ would be a reasonable estimate. The age of the Universe is discussed in terms of the parameter values.

## 1 Introduction

In this paper I summarize the observational determination of the cosmological parameters and give my personal prospect for the 21st century.

We are living in the expanding Universe. Because of this cosmological expansion, distant galaxies are all receding away from us. This recession velocity is measured by the Doppler displacement of spectral lines in the spectra of galaxies. We define the *redshift* $z$ by

$$z = (\lambda_{obs} - \lambda_0)/\lambda_0, \tag{1}$$

where $\lambda_{obs}$ is the observed wavelength of a spectral line in the spectrum of a galaxy and $\lambda_0$ is the rest frame wavelength of the line. The recession velocity is computed by

$$v = c\frac{(1+z)^2 - 1}{(1+z)^2 + 1}, \tag{2}$$

where $c$ is the velocity of light. In the nearby universe where $z << 1$, we have

$$v \sim cz. \tag{3}$$

What Hubble found in 1929 is that the ressession velocity is proportional to the distance of galaxies (Hubble 1929; 1936). This is called the Hubble's law.

The standard model to describe the expanding Universe is the Friedmann models (Friedmann 1924). According to the Friedmann models, structure and time evolution of the Universe are characterized by two parameters, the Hubble constant $H_0$ and the dimension-less density parameter $\Omega_0$. The Hubble constant represents the present expansion rate of the Universe. The density parameter represents the present mean density of the Universe in units of the critical density $\Omega_{0,c} = 3H_0^2/8\pi G$, which is the mean density of a universe where the expansion energy is equal to the gravitational potential energy.

Figure 1 shows the behavior of the *scale factor* of Friedmann models as a function of time for different values of $\Omega_0$. The scale factor $a(t)$ is a convention to express the cosmological expansion. The coordinate of a point $\mathbf{r}(t)$ in the expanding Universe is written as $\mathbf{r}(t)=a(t) \times \boldsymbol{\xi}$ using the time independent proper coordinate $\boldsymbol{\xi}$. If $0 < \Omega_0 \leq 1$, the Universe continues to expand forever. On the other hand, if $\Omega_0 > 1$, the Universe will stop the expansion some time in the future and start contracting. The Hubble constant is the delivative of the lines in Fig. 1 at the present epoch, *i.e.*, $t = T_0$, and it determines the order of the age of the Universe. The actual age of the Universe is modulated

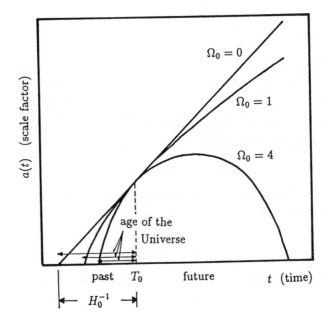

Fig. 1 Behavior of the scale factor $a(t)$ of Friedmann models with different values of $\Omega_0$.

by $\Omega_0$ as

$$T_0 = H_0^{-1} f(\Omega_0), \tag{4}$$

as shown in Fig. 2, where $f(\Omega_0) = 1$ for $\Omega_0 = 0$ and $f(\Omega_0) = 2/3$ for $\Omega_0 = 1$.

There could be another parameter, the cosmological constant $\Lambda$, in the Friedmann models. There is, however, little convincing observational evidence for the non-zero $\Lambda$ at present. Accordingly, I assume $\Lambda = 0$ in this paper.

age (in units of $H_0^{-1}$)     ($\Lambda = 0$)

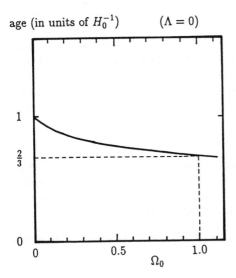

Fig. 2 The age of a Friedmann model (in units of $H_0^{-1}$) as a function of $\Omega_0$.

## 2   Determination of the Hubble Constant $H_0$

### 2.1   Method

From the observational point of view, the Hubble constant is the constant of proportionality in the Hubble's law

$$v_H = H_0\, r, \tag{5}$$

where $v_H$ is the recession velocity due to the cosmic expansion and $r$ is the distance of a galaxy. The observed line of sight recession velocity of a galaxy, however, consists of three components as

$$v_{obs} = v_H + v_{bulk} + v_{rand}, \tag{6}$$

where $v_{bulk}$ and $v_{rand}$ are the line-of-sight components of the so-called bulk motion velocity and the random motion velocity, respectively. The velocity components other than $v_H$ are called peculiar velocity. The bulk motion is the systematic motion of galaxies seen, for example, near a big cluster of galaxies. If we take a small volume at some distance from the cluster, all the galaxies in the volume move toward the cluster more or less in concert due to the gravitational pull of the cluster.

The random velocity of galaxies is estimated to be $\lesssim 200$ kms$^{-1}$ over a scale of $\lesssim 20$ Mpc (Aaronson *et al.* 1982). We do not know yet, however, how large is the bulk velocity and how widely the bulk motion shows coherence (see Strauss and Willick 1996 for a review). There is possibility that measurement based on galaxies in a small volume gives a biased result due to the peculiar velocity. The Local Group of galaxies, which includes our Galaxy, is known to be moving with respect to the cosmic microwave background radiation at $\sim 600$kms$^{-1}$ (Smoot *et al.* 1991). However, since $v_H$ is proportional to the distance, $v_H \gg (v_{bulk} + v_{rand})$ for distant galaxies. So, we expect to obtain the Hubble constant by

$$H_0 = <v_{obs}/r>, \tag{7}$$

for a sample of reasonably distant galaxies, where $<>$ means the sample average. The derivation of $v_{obs}$ is straightforward once the spectrum of a galaxy is obtained. There are, however, many problems to measure the distance $r$ of galaxies. The problem of the determination of the Hubble constant has been, in fact, the problem of measuring galaxy distances.

### 2.2 Principles of Measuring Galaxy Distances

The conventional principle to measure galaxy distances is to use *standard candles* as distance indicators. Suppose that we observe the flux $f$ of an object with the luminosity $L$ located at a distance $r$ and compare the flux with the reference flux $f_{10}$ observed at the distance of 10 parsec (pc; 1pc=$3\times10^{13}$km), we have

$$f = L/4\pi r^2, \quad f_{10} = L/4\pi 10^2, \tag{8}$$

and

$$f/f_{10} = (10/r)^2. \tag{9}$$

The magnitude is the favorite unit used in astronomy to measure the flux. The *apparent* magnitude is defined by

$$m = -2.5\log_{10}f + C, \tag{10}$$

while

$$M = -2.5\log_{10}f_{10} + C, \tag{11}$$

is called the *absolute* magnitude, where the constant term $C$ defines the zero point of the magnitude scale. From equations (9), (10), and (11) we have

$$m - M = 5\log_{10} r - 5 + A. \tag{12}$$

The interstellar space is not completely transparent and the last term $A$ represents the correction for the interstellar absorption. The quantity

$$(m - M)_0 = m - M - A, \tag{13}$$

is called the *distance modulus*.

The standard candle is an object whose intrinsic brightness, hence the absolute magnitude, is known by some means. Then, the distance of the standard candle is obtained by measuring the apparent magnitude and the interstellar absorption through the formula

$$\log_{10} r = 0.2(m - M)_0 + 1, \tag{14}$$

We use different standard candles for different distance ranges; brighter candles for more distant galaxies. Measuring galaxy distances step by step toward more distant galaxies using different standard candles is often called *the cosmological distance ladder* (Rowan-Robinson 1985).

There are other methods based on different principles to measure the distance of galaxies or clusters of galaxies which I will describe later. They include the expanding photosphere method for supernovae type II, Sunyaev-Zel'dovich effect, and the time delay in gravitationally lensed images.

One important remark on the standard candles is that their physical basis is not completely understood. The most reliable, *i.e.*, the best understood standard candle is the Cepheid variable stars, which is called the primary distance indicator. The error in the Cepheid distance can be made as small as $\lesssim \pm 10\%$ (Madore anf Freedman 1991; Jacoby *et al.* 1992). Other standard candles, which are called the secondary distance indicators, are less reliable than Cepheids. The distance given by these secondary indicators usually have the *internal error* of 10–15% level. However, the *systematic error* is not known very well although comparisons of distances of the same galaxies/clusters by different indicators give some estimates of the systematic errors (Jacoby *et al.* 1992). Accordingly, we use many different candles and/or different methods based on different principles to avoid the effect of unknown systematic errors: 'do not put all the eggs in a single basket.'

### 2.3 Standard Candles as Distance Indicators

(a) Primary Distance Indicator: Cepheids

Cepheids, the primary distance indicator, are bright variable stars which exhibit the characteristic light curve as shown in Fig. 3. The variabiliy is due to the pulsation of the star itself. It is known that there is a strong correlation between the period of variation and the absolute magnitude, which is called the period-luminosity relation (*e.g.*, Feast and Walker 1987); Cepheids with longer variation periods have brighter absolute magnitudes. Then, once we identify a Cepheid in a distant galaxy and observe the apparent magnitude and the period of variation, we can estimate the absolute magnitude and obtain the distance from equation (14). In practice, observation is usually made at more than two wavelengths to estimate the interstellar absorption and for many Cepheids to increase the statistical accuracy. Cepheids could be observed using ground-based telescopes only up to 4 Mpc (1Mpc=$10^6$ pc) until recently with few exceptions (Pierce *et al.* 1994). However, the refurbished Hubble Space Telescope (hereafter HST) can now observe Cepheids in a galaxy as distant as 20 Mpc (Freedman *et al.* 1994; Saha *et al.* 1996; Sandage *et al.* 1996).

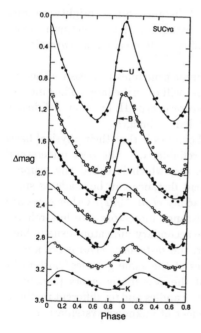

Fig. 3 Light curves of a typical Cepheid at various wavelengths; shorter to longer wavelengths from top to bottom. $U \sim 0.35\mu m$, $B$, $V$, and $R \sim 0.7\mu m$ stand for ultraviolet, blue, visual, and red bands, respectively in the optical wavelength region. $I$, $J$, and $K \sim 2.2\mu m$ denote the near infrared bands. (taken from Madore and Freedman 1991).

(b) Secondary Distance Indicators: PNLF, SBF, and DIRs

It is beyond the scope of this paper to give detailed explanations to many secondary distance indicators (see Jacoby *et al.* 1992 for a review). I will only introduce some representative ones which are believed to have good accuracies.

The PNLF method is based on the Planetary Nebulae Lumoinosity Function and uses its brightest end as a standard candle (Jacoby *et al.* 1990). A star with mass in the range of $0.8-7m_\odot$ ($m_\odot$ is the mass of the sun) ejects its outer envelope at the late phase of evolution. The ejected gas reprocesses the ultraviolet radiation energy from the hot central star (core of the original star) to emit emission lines in the visible wavelength region. This bright gaseous shell is the planetary nebula. Model computations show that the mass range of the central star which develops a bright planetary nebula is quite narrow (Jacoby 1989) and that the brightness of the planetary nebula is only weakly dependent on the metallicity (Jacoby *et al.* 1992). This is a feature that is preferable for a standard candle.

The SBF method makes use of the Surface Brightness Fluctuation of a galaxy as distance indicators (Tonry and Schneider 1988). Consider a galaxy which is composed of a single kind of stars with identical absolute magnitude. The surface brightness is the sum of the fluxes from the stars that are seen in a unit angular area, for example, one square arcsecond. For the mean surface brightness of $I = Nf$, we expect the fluctuation $\Delta I = N^{1/2}f$ due to the star number statistics obeying the Poisson distribution, where $N$ is the mean number of stars seen in the unit area and $f$ is the flux of a star. We assume that the errors other than that due to the star number fluctuation such as photon noise, interstellar absorption, etc. are negligibly small. Then, the ratio of the surface brightness fluctuation squared to the mean surface brightness scales to the flux as

$$\Delta I^2/I = f. \tag{15}$$

Accordingly, we can determine the galaxy distance by measuring the ratio $\Delta I^2/I$ if appropriate calibration is made for $f$ (Tonry 1991). In practice, the surface brightness fluctuation is measure by the power spectra using the Fourier transform of the galaxy image (*e.g.*, Pahre and Mould 1994).

The brightest available standard candle is, of course, a galaxy itself. However, galaxies exist in the absolute magnitude range more than a factor of 100. Accordingly, in order to use a whole galaxy as a standard candle we need some means to estimate its absolute magnitude. Distance Indicator Relations (DIRs) provide the means. A DIR is an empirical correlation between distance-dependent observables and distance-independent observables of galaxies. The former includes apparent magnitude and apparent size and the latter includes internal velocity and color of galaxies. In the case of the Tully-Fisher relation for spiral galaxies (Tully and Fisher 1977) illustrated in Fig. 4, the former is the magnitude (corrected for interstellar absorption) and the latter is the rotational velocity measured by the velocity width of the $\lambda 21$ cm emission line originated from the neutral hydrogen gas in the galaxy. We first establish the correlation using so-called *Local Calibrators*, which are nearby galaxies whose distances are measured by the primary distance indicator. Next we observe the same correlation for galaxies in a distant cluster of galaxies, which can be regarded to be at the same distance. Then as shown schematically in Fig. 4, the shift of the two relations along the ordinate, *i.e.*, the magnitude axis, gives us the distance modulus $(m - M)_0$ of the cluster.

One important but *implicit* assumption in this method is that the DIR is the same everywhere in the Universe. However, galaxy formation is a process which we do not understand yet, and there is no physical basis at present for the assumption that galaxies in the vicinity of our Galaxy follows the same correlation as galaxies in the different part of the Universe where the galaxy environments may well be different. The universality of the DIRs is an important issue to be confirmed by future observations. Difficulty lies in the fact that there are few nearby clusters of galaxies that represent different environments in the Universe. In case of only a few nearby clusters such as Virgo cluster and Ursa Major cluster, we can observe significant fraction of galaxies down to a reasonably faint absolute magnitude and examine the correlation over a wide

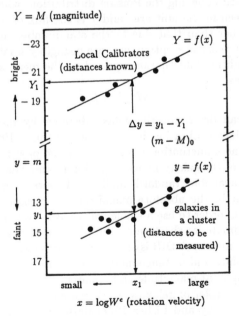

Fig. 4 Schematic Tully-Fisher relation.

magnitude range. In case of more distant clusters, however, because of the limiting magnitude, *i.e.*, the limit of observation imposed on the apparent magnitude, we can observe intrinsically brighter galaxies which is a smaller fraction of the total galaxy popullation. Figure 5 illustrates the difficulty by showing the limiting magnitudes of the Tully-Fisher relation for different clusters (Kraan-Korteweg *et al.* 1988) translated into the Virgo cluster.

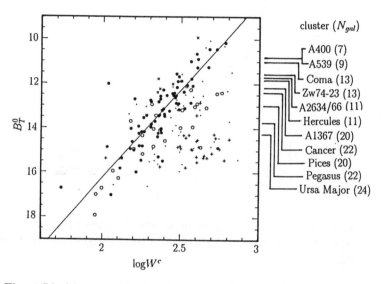

Fig. 5 Limiting magnitudes for several clusters of galaxies translated to the Virgo cluster distance and shown superposed on the Tully-Fisher relation of the Virgo Cluster taken from Yasuda *et al.* (1997). Filled and open circles denote 'certain' and 'possible' cluster members. Pluses are 'background' galaxies and crosses are 'HI deficient' galaxies. Dots are the data with uncertain accuracies. Number of galaxies available to the analysis is shown in the parentheses.

## (c) Supernovae Type Ia (SNe Ia)

Supernovae type Ia are an important standard candle. A SN Ia is the thermonuclear explosion of a white dwarf in a binary system caused by mass accretion from a companion star. The observed characteristics of SNe Ia are fairly homogeneous and a typical light curve is shown in Fig. 6. The absolute magnitude at the peak brightness, $M^{max}$, is believed to be nearly constant since the mass of the exploding white dwarf is usually assumed to be at the

164

Chandrasekhar limit of $1.4 m_\odot$. The observational homogeneity of SNe Ia has been often considered to imply their origin from a single progenitor system. There is, however, possibility that a variety of progenitor systems are contributing (Branch *et al.* 1995).

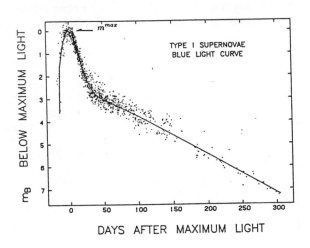

DAYS AFTER MAXIMUM LIGHT

Fig. 6 Composite light curves of 38 SNe Ia made with the compilation by Barbon *et al.*(1973) (taken from Doggett and Branch 1985).

The importance of SNe Ia as a standard candle lies in the fact that they are almost as bright as a galaxy. We can go very deep into the Universe with SNe Ia in spite of the disadvantage that we cannot forsee and control their occurence in a particular galaxy. For the peak brightness of SNe Ia, using equations (5) and (12) and applying an appropriate correction for interstellar absorption, we have

$$m^{max} = 5 \log_{10} v_H + M^{max} - 5 \log_{10} H_0 - 5, \qquad (16)$$

where $v_H$ is the recession velocity of the SN Ia host galaxy. If $M^{max}$ is nearly constant among SNe Ia, the plot of $m^{max}$ versus $\log_{10} v_H (\sim \log_{10} v_{obs})$ must have a slope of 5 and will show a small dispersion. This is what actually observed as shown in Fig. 7 (Tammann and Sandage 1995). The intercept of the diagram determines $H_0$, *if $M^{max}$ is known*. The intercept is a quantity which is rather well defined by many SNe Ia observed so far. Accordingly, how to calibrate $M_{max}$ is the problem of $H_0$ determination using SNe Ia. There were no nearby

SN Ia host galaxies whose distance could be measured with Cepheids before the HST. However, the HST key project, 'extragalactic distance scale', is now under way in which the distance of some 25 galaxies will be determined using Cepheids. The galaxies selected are those which give the calibration of various secondary distance indicators. The Cepheid distance was available for only two SN Ia host galaxies in 1994 (Sandage *et al.* 1994), but the number has increased to six now (Sandage *et al.* 1996; Schaefer 1996). There is, however, still controversy among SN Ia specialists on how homogeneous SNe Ia are and how to define a good sample of SN Ia by rejecting *peculiar* ones (Branch *et al.* 1993) or those with *bad* photometry (Hamuy *et al.* 1995). Which are peculiar and which photometry is bad depends more or less on authors. It has become widely accepted that there is an intrinsic dispersion in $M^{max}$ and that brighter SNe Ia fades more slowly (Phillips 1993; Riess *et al.* 1995). However, how much this effect actually afffect the analysis again depends on the sample selection, that is, on authors.

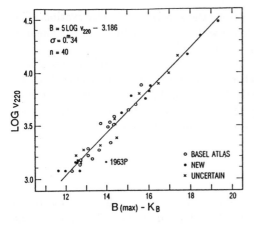

Fig. 7 The peak apparent magnitude of SNe Ia in the $B$ band as a function of the recession velocity of their host galaxies (taken from Tamann and Sandage 1995).

## 2.4 Expanding Photosphere Method for Supernovae Type II (SNe II)

This is a method to measure the parallax of the expanding photosphere of a supernova type II (SN II), which is an explosion of a massive star (Schmidt *et al.* 1992). Suppose a SN II with a photosphere of radius $R$ located at the distance $r$. The flux from this SN II can be written as

$$f_\nu = \frac{\pi R^2 B_\nu(T)\zeta^2(T)}{r^2},\qquad(17)$$

where $B_\nu(T)$ is the Planck function and $\zeta^2(T)$ is the correction factor for the departure of the actual SN II radiation from the black body radiation, *i.e.*, $\zeta^2(T)=1$ for the black body radiation. The parallax $\theta$ can be measured not directly but by using observable quantities and $\zeta^2(T)$ as

$$\theta = \frac{R}{r} = \left(\frac{f_\nu}{\pi B_\nu(T)\zeta^2(T)}\right)^{1/2}, \tag{18}$$

where $f_\nu$ and $T$ are obtained from multi-color photometric observation and $\zeta^2(T)$ is estimated from a model. We assume free expansion for the photosphere as

$$R = v(t - t_0) + R_0, \tag{19}$$

where $t_0$ is the time of the explosion and $R_0 = R(t_0)$. Assuming $R \gg R_0$, we have

$$t = r\left(\frac{\theta}{v}\right) + t_0. \tag{20}$$

The expansion velocity $v$ is obtained from the spectroscopic observation. The plot of $t$ as a function of $(\theta/v)$ is then nearly a straight line and the slope of the line gives the distance $r$ and the intercept gives $t_0$.

The parameter $\zeta^2(T)$ is estimated by a model. Accordingly, the reliability of this method depends on how well the model can represent the real SN II. This was tested for the SN 1987a in the Large Magellanic Cloud whose distance is known by the primary indicators (Schmidt *et al.* 1992). However, it is not straightforward to construct a good model for each of different SNe II (Schmidt *et al.* 1994).

### 2.5 Sunyaev-Zel'dovich Effect

X-ray observations have revealed that most clusters of galaxies contain hot plasma with temperature of $\sim 10^7$K (David *et al.* 1993). The hot plasma interacts with the cosmic microwave background (CMB) radiation through inverse Compton scattering, causing the distortion of the CMB spectrum. The distortion is observed as the temperature decrement $\Delta T$ of the CMB radiation in the Rayleigh-Jeans region, *i.,e.*, radio wavelength region (Sunyaev and Zel'dovich 1972). The temperature decrement is writen as

$$\Delta T \propto n_e T_e L, \tag{21}$$

where $n_e$ and $T_e$ are the electron densisty and the electron temperature of the plasma and $L$ is the the extension of the plasma along the line of sight. On the other hand, the X-ray luminosity $S_X$ of the plasma is written as

$$S_X \propto \frac{n_e^2 T_e^{1/2} L^3}{r_L^2}, \tag{22}$$

where $r_L$ is the luminosity distance. The spherical symmetry is assumed for the plasma distribution and

$$L = \theta r_A, \tag{23}$$

where $\theta$ is the angular size of the plasma distribution measured by the X-ray observation and $r_A$ is the angular diameter distance. Eliminating $n_e$ from equations (21) and (22) we have

$$\frac{r_L^2}{r_A} \propto (\frac{\theta}{T_e^{3/2}})(\frac{\Delta T^2}{S_X}). \tag{24}$$

The left-hand side is a function of the cosmological parameters $H_0$ and $\Omega_0$, and it can be computed by the observable quantities in the right-hand side. It is noted that $\Omega_0$ is also involved in the computation since most clusters are distant and we cannot use the approximation $z \ll 1$ (Kobayashi et al. 1996).

The determination of $H_0$ by this method needs no recource to the cosmological distance ladder. The physical basis of the phenomenon is also well understood. However, there are only a few high-quality determinations so far made because of the technical difficulties in the observations, especially in the radio observations. The advent of the synthesis radio telescope available to the measurement of the temperature decrement, however, has resulted in a much improved accuracy (Jones et al. 1993). The uncertainty in the X-ray model will be the future major uncertainty. This is a promising method to measure $H_0$ in the distant Universe and there is a big hope in future.

## 2.6  Time Delay of the Gravitationally Lensed Images

Finally I describe a method to measure $H_0$ using the time delay of gravitationally lensed images. Figure 8 illustrates the situation schematically. The light emitted from a quasar behind a cluster of galaxies is deflected by the mass of the cluster, and the observer sees two images A and B of the quasar. Suppose that the quasar suddenly brightens. This brightening is observed both in image A and in image B but at different epochs because the light paths for the two images are different. The time delay $\Delta\tau$ is expressed by

$$\Delta\tau = -\epsilon(1+z_L)\frac{r_L r_S}{r_{LS}}(\theta_A - \theta_B)\frac{\alpha_A + \alpha_B}{2}, \tag{25}$$

where the subscriptes A and B are for image A and image B, and L and S are for lens (cluster) and source (quasar), respectively, $0 < \epsilon \leq 1$ is a parameter representing the dependence on the mass model of the lens, and $z_L$ is the resdshift of the lens (Falco *et al.* 1991). The term $r_L r_S / r_{LS}$ contains $H_0$ (and $\Omega_0$) and all the other quantities in equation (25) are constrained from observations. There are only a few determinations so far made with this method (Kundić *et al.* 1997). This is also a promising method to measure $H_0$ in the distant Universe.

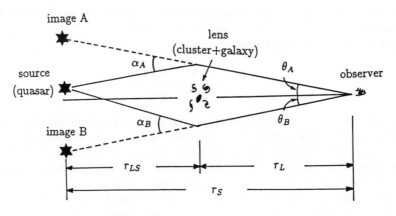

Fig. 8 Schematic view of a gravitationally lensed images.

## 2.7 Summary of $H_0$ Determinations

Figure 9 gives the state-of-the-art summary of $H_0$ determinations by various methods I described. For the sake of brevity, I have tried to select one or two representative *values*, *i.e.*, not the extreme values at either the large or small sides among the determinations by different authors by the same method. The perpendicular bars in Fig. 9 show the error bars for the $H_0$ values while the horizontal bars show the distance coverage of the respective methods. Cepheids and PNLF cover the space within about 20 Mpc and SBF covers up to 50 Mpc. DIRs reach about 100 Mpc. SNe II reach 200 Mpc and SNe Ia reach up to 400 Mpc. The broken horizontal lines for the Cepheids indicates that while the Cepheid observations were made at $\lesssim 15$ Mpc the $H_0$ value was derived on the basis of the Coma cluster distance calibrated by the Cepheid observations. It should be noted that the Sunyaev-Zel'dovich effect and the gravitational lens method can measure $H_0$ at much farther distances than other methods.

The interim report of the HST key project for the cosmological distance scale give $H_0 = 73\pm10$ (Mould 1996), which is consistent with most determinations within 100 Mpc shown in Fig. 9. Only a discrepant value is $H_0 = 57\pm5$ by the SN Ia method. However, the error bar for this measurement represents the internal error only and no allowance is made for possible systematic effects. I conclude from Fig. 9 that $H_0 \sim 70 \pm 10$ is the best estimate of the local ($r \lesssim 100$ Mpc) value of the Hubble constant. For convenience of quotation, I will take the value by Mould (1996). There is a hint in Fig. 9 that the far field ($r > 100$ Mpc) value is slightly smaller than the local value. However, this should be confirmed by future measurements using more objects. There is good hope in SN II, SN Ia, Sunyaev-Zel'dovich effect, and gravitational lens methods.

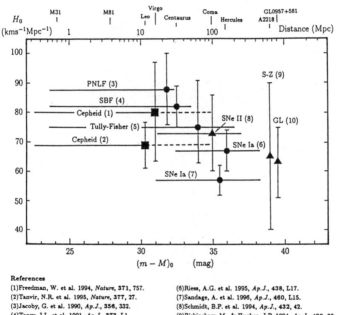

**References**

(1)Freedman, W. et al. 1994, *Nature*, **371**, 757.

(2)Tanvir, N.R. et al. 1995, *Nature*, **377**, 27.

(3)Jacoby, G. et al. 1990, *Ap.J.*, **356**, 332.

(4)Tonry, J.L. et al. 1991, *Ap.J.*, **373**, L1.

(5)Watanabe, M. 1996, *PhD thesis*, University of Tokyo.

(6)Riess, A.G. et al. 1995, *Ap.J.*, **438**, L17.

(7)Sandage, A. et al. 1996, *Ap.J.*, **460**, L15.

(8)Schmidt, B.P. et al. 1994, *Ap.J.*, **432**, 42.

(9)Birkinshaw, M., & Hughes, J.P. 1994, *Ap.J.*, **420**, 33.

(10)Kundić, T. et al. 1997, submitted to *Ap.J.*

Fig. 9 Summary of recent $H_0$ determinations by various methods explained in the text. References are shown by the numbers in the parentheses. The ordinate is the value of $H_0$ and the abscissa is the distance. Locations of some representative galaxies and clusters of galaxies are shown in the top margin. Perpendicular error bars show the error of $H_0$ quoted in the references. Horizontal bars indicate the approximate distance coverage of the respective methods.

## 3   Hubble Constant and the Large Scale Structure of the Universe

A great advance was made in late 1980's in the study of the three dimensional distribution of galaxies in the Universe by extensive *redshift surveys*. In a redshift survey, we obtain redshifts of all the galaxies in the survey area which are brighter than a certain limiting magnitude. Redshifts or recession velocities can be used as a measure of the distance (equations (3) and (5)). Accordingly, with the redshift and the two-dimensional position on the sky, we can map the

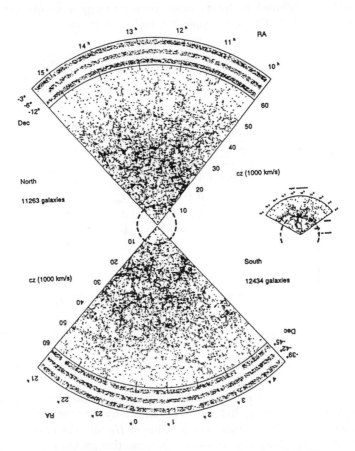

Fig.   10 Wedge diagrams constructed from the CfA-II survey (Geller *et al.*1987; small wedge) and from the Las Campanas Survey (Shectman *et al.*1996; two large wedges). The circles shown by the broken line has the radius of about 100 Mpc.

three-dimensional distribution of galaxies, which are often shown for a slice of the Universe in the fan-shaped *wedge diagram*. Figure 10 shows the wedge diagrams obtained by the Center for Astrophysics (CfA) survey (Geller *et al.* 1987) and by the Las Campanas Survey (Shectman *et al.* 1996). Our galaxy is located at the pivot of the fan and the observed radial velocity ($v_{obs} = cz$) is plotted along the radius of the fan. Each dot in the diagram represent a galaxy. As seen in the CfA survey diagram, there are superclusters of galaxies that consists of a few clusters connected by wall-like structure and voids where few galaxies are found. The scale of such features exceeds $5000 \text{kms}^{-1} \sim 70 \text{Mpc}$, which is much larger than the typical size of clusters of galaxies (5 Mpc) that had been believed to be the largest structure in the Universe. This large scale structure of the Universe composed of superclusters and voids are found to extend up to $\sim 60000 \text{ kms}^{-1}$ as seen in the Las Campanas survey.

We measured $H_0$ by several methods applied to many galaxies only within $r \lesssim 100$ Mpc as shown in the preceding section. The size of this region is comparable to the size of the large scale structure. Accordingly, there is possibility that the present local measurement of $H_0$ is biased by peculiar velocities due to the large scale structure. Difference between local measurements and the global value is investigated, for example, by Nakamura and Suto (1996).

## 4   Determination of the Density Parameter $\Omega_0$

The density parameter $\Omega_0$ is much less constrained by observations than $H_0$. I will briefly describe the methods to determine $\Omega_0$ below, but significant determinations are the prospect for the 21st century.

### 4.1   Dynamical Test

The density parameter can be obtained by the analysis of peculiar velocities of galaxies. Suppose that there is a galaxy at a distance $r$ from a cluster of galaxies. The peculiar velocity $v_p$ of the galaxy due to the gravitational pull of the cluster is expressed by

$$\frac{v_p}{H_0 r} \sim \frac{1}{3}\Omega_0^{0.6}\frac{\delta\rho}{\rho}, \qquad (26)$$

where $\rho_0$ is the average density of the Universe and $\delta\rho$ is the density enhancement within the cluster (Davis and Peebles 1983). In the actual analysis, $\Omega_0$ can be derived without knowing the value of $H_0$. However, the method does require *relative* distances to galaxies in order to obtain $v_p = v_{obs} - H_0 r$. As we already saw, the galaxy distances are difficult quantities to measure. Accordingly, $v_p$ is even more difficult quantity to obtain from the observation.

Another difficulty lies in the density contrast $\delta\rho/\rho$, which is measured by the distribution of galaxies. How to account for the dark matter content in the density contrast is an open question. There is a good review on the peculiar velocity measurements by Strauss and Willick (1996).

## 4.2 Structure Formation

The evolution of the large scale structure as a function of the redshift is known to be dependent on $\Omega_0$ (e.g., Peacock and Dodds 1996). Accordingly, a quantitative comparison of the large scale structure at present epoch ($z \lesssim 0.2$) with that in the past ($z >> 0.2$) will give us constraints on $\Omega_0$. However, available data are not sufficient either for the present epoch or for the past to give significant constraints. The data for the present epoch will be enhanced by the Sloan Digital Sky Survey to be completed in the beginning of the 21st century (Gunn and Knapp 1993; Okamura 1995; Gunn and Weinberg 1995). The data for the past have been accumulated by the Keck 10m telescopes and will also be collected by the several 8m telescopes under construction including the Japanese SUBARU telescope (Kaifu 1996; Iye et al. 1996; Nishimura et al. 1996: Iye 1997).

## 4.3 Geometrical Tests

In the geometrical tests one tries to measure the space curvature due to the mass of the Universe by looking at very distant objects (e.g., Sandage 1988). There are several tests based on different observations: the magnitude-redshift relation, which is often called the Hubble diagram, for standard candles such as the brightest galaxy in clusters (Kristian et al. 1978; Yoshii and Takahara 1988), the size-redshift relation for standard rods such as the core size of radio sources (Kellerman 1993), and number-magnitude relation (galaxy count) for field galaxies (Djorgovski et al. 1995).

There is, however, an essential difficulty common to all the geometrical tests. It is the evolution of standard candles. Nothing remains constant standard candles in the whole history of the evolving Universe. And the evolutionary effects couple with the geometry effect. One example is shown in Fig. 11, which is taken from Yoshii and Takahara (1988). Figure 11 shows the Hubble diagram for the brightest cluster galaxies. Three thin lines shows the geometry effects for $q_0 \equiv \Omega_0/2 =0.02$, 0.5, and 1 in case that there is no evolution in the brightness of such galaxies. It is clear that the no-evolution models do not fit to the observations. Four thick lines represent the predictions of evolution models in which the galaxies are assumed to have been born at $z =5$ (three lines for $q_0 \equiv \Omega_0/2=0.02$, 0.5, and 1) and at $z =3$ (for $q_0 \equiv \Omega_0/2 = 1$)

and flushed up at birth and decreased their brightness according to the star formation activity taking place in the galaxy.

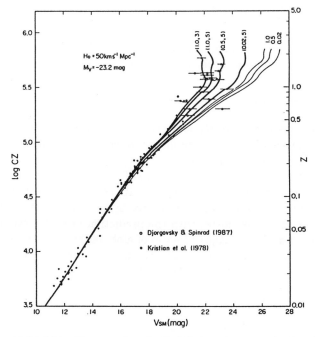

Fig. 11 Hubble diagram for the brightest cluster galaxies (taken from Yoshii and Takahara 1988).

Fig. 11 shows that the evolutionary effects are much larger than the geometry effect in $z \lesssim 2$ and that it is difficult to draw definite conclusion on the value of $\Omega_0$ once the standard candles are known to evolve. Accordingly, evolutionary history of standard candles, or more generally speaking, evolutionary histotry of galaxies should be elucidated before we can obtain constraints on $\Omega_0$ from the geometrical tests. Recently, HST presented unprecedentedly deep ($I \sim 28$ mag) images of distant galaxies in the Hubble Deep Field (HDF; Williams *et al.* 1996). The morphological type mix of HDF galaxies is found to be quite different from that of nearby galaxies; population of elliptical/S0 galaxies looks similar but there are more irregular/merging galaxies (van den Bergh *et al.* 1996). We must be seeing young galaxies soon after their birth in the HDF. Systematic spectroscopy of HDF galaxies and galaxies of similar magnitudes in other fields, which takes a long time even with the 8-10m telescopes, will eventually tell us the evolutionary history of galaxies (*e.g.*, Koo *et al.* 1996; Cowie *et al.* 1996).

## *4.4   Summary of $\Omega_0$ Determinations*

As we have seen above, $\Omega_0$ is not well constrained by the observation at present. I propose here a conservative range of $0.1 < \Omega_0 \leq 1$ as a basis for the following discussion. The inflation theory predicts $\Omega_0$=1 (Sato 1981; Guth 1981).

## 5   The Age Problem

Having the constraints on $H_0$ and $\Omega_0$, we can now discuss the age of the Universe. The age of the Universe should, of course, be longer than the age of the oldest objects in the Universe.

Globular clusters residing in the halo of our Galaxy are one of the oldest objects. A globular cluster consists of $10^5 - 10^6$ stars that were born at the same epoch. The age of a cluster can be estimated by comparing the color-magnitude diagram of stars in the cluster with the prediction of stellar evolution theory (*e.g.*, Hesser *et al.* 1987). Age estimates of globular clusters are in the range of 12–22 Gyr (*e.g.*, Carney *et al.* 1992). Taking the various errors into account, Chaboyer *et al.* (1996) derived 12 Gyr as the *lower limit* of the age of 17 oldest globular clusters.

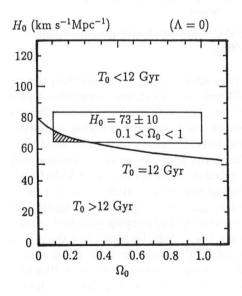

Fig. 12 The age problem. The thick solid curve shows the values of $(H_0, \Omega_0)$ that give the age $T_0$=12 Gyr, which is the lower limit of the age of Galactic globular clusters. The observational constraints on the $H_0$ and $\Omega_0$ are shown by the rectangle. The allowed region is shown hatched.

Figure 12 illustrates the so-called age problem in the parameter space of $(\Omega_0, H_0)$. The solid curve represent the parameter values that give the age of the Universe of 12 Gyr (equation (4)). The observational constraints on the parameters, $H_0 = 73 \pm 10$ kms$^{-1}$Mpc$^{-1}$ and $0.1 < \Omega_0 \le 1$, are shown by the rectangle. The constraint that the age of the Universe should be longer than 12 Gyr gives the following conclusions.

1. There is an allowed region (shown hatched), though very small, in the parameter space. The local value of $H_0 = 73 \pm 10$ is consistent with the globular cluster age limit *only for* $\Omega_0 \lesssim 0.3$.

2. On the other hand, if we stick to $\Omega_0 = 1$, the globular cluster age limit is consistent *only for* $H_0 \lesssim 54$, which is outside of the $\sim 1\sigma$ error of the local value. There is some hint that the far field ($r > 100$ Mpc) value of $H_0$ is smaller than the local ($r \lesssim 100$ Mpc) value. However, this is an open question at present and should be checked by future observations.

I do not think that there is an immediate need for the cosmological constant because of the age problem shown in Fig. 12. It should, however, be invoked if future observations confirm the discrepancy by giving a far field value of $H_0$ and/or by giving a more accurate estimate of $\Omega_0$.

## 6  Concluding Remarks

We cannot answer yet the basic questions about the Universe in which we are living. How old and how large is it? What is its ultimate fate? Is there non-baryonic dark matter? Does the non-zero cosmological constant exist?, etc, etc.

This is because we have not observed a large volume enough to represent the whole Universe and because we have not observed the formation epoch and early evolutionary stage of galaxies. Extensive photometric/redshift surveys are under way and deep spectroscopy of very faint galaxies with 8-10m telescopes is within our reach. There is good hope to answer the questions in the early 21st century.

## References

Aaronson, M., Huchra, J., Mould, J., Schechter, P.L., and Tully, R.B. 1982, *Astrophys. J.*, **258**, 64.

Barbon, R., Ciatti, F., and Rosino, L. 1973, *Astron. Astrophys.*, **25**, 241.

Birkinshaw, M., and Hughes, J.P. 1994, *Astrophys. J.*, **420**, 33.

Branch, D., Fisher, A., and Nugent, P. 1993, *Astron. J.*, **106**, 2382.

Branch, D., Livio, M., Yungelson, L.R., Boffi, F.R., and Baron, E. 1995, *Publ. Astron. Soc. Pacific*, **107**, 1019.

Carney, B.W., Storm, J., and Jones, R.V. 1992, *Astrophys. J.*, **386**, 663.

Chaboyer, B., Demarque, P., Kernan, P.J., and Krauss, L.M. 1996, *Science*, **271**, 957.

Cowie, L.L., Songaila, A., Hu, E.M., and Cohen, J.G. 1996, *Astron. J.*, **112**, 839.

David, L.P., Slyz, A., Jones, C., Forman, W., and Vrtilek, S.D. 1993, *Astrophys. J.*, **412**, 479.

Davis, M., and Peebles, P.J.E. 1983, *Ann. Rev. Astron. Astrophys.*, **21**, 109.

Djorgovski *et al.* 1995, *Astrophys. J.*, **438**, L13.

Doggett, J.B., and Branch, D. 1985, *Astron. J.*, **90**, 2303.

Falco, E.E., Gorenstein, M.V., and Shapiro, I.I. 1991, *Astrophys. J.*, **372**, 364.

Feast, M.W., and Walker, A.R. 1987, *Ann. Rev. Astron. Astrophys.*, **25**, 345.

Freedman, W.L. *et al.* 1994, *Nature*, **371**, 757.

Friedmann, A. 1924, *Z. Phys.*, **10**, 377, and *Z. Phys.*, **21**, 326.

Geller, M.J., Huchra, J.P., and de Lapparent, V. 1987, *IAU Symp.*, **124**, 301.

Gunn, J.E., and Knapp, G.R. 1993, *ASP Conf. Ser.*, Vol.**43**, ed. B.T. Soifer (Provo: Brigham Young Univ.), P263.

Gunn, J.E., and Weinberg, D. 1995, in *Wide Field Spectroscopy and the Distant Universe*, eds. Maddox, S.J., and Aragón-Salamanca (Singapore: World Scientific Publ. Co. Ltd), p.3.

Guth, A.H. 1981, Phys. Rev, **D23**, 347.

Hamuy, M., Phillips, M.M., Maza, J., Huchra, J., Schommer, R.A., and Cornell, M.E. 1995, *Astron. J.*, **109**, 1.

Hesser, J.E., Harris, W.E., vandenBerg, D.A., Allwright, J.W.B., Shott, P., and Stetson, P.B. 1987, *Publ. Astron. Soc. Pacific*, **99**, 739.

Hubble, E. 1929, *Proc. Nat. Acad. Scie.*, **15**, 168.

Hubble, E. 1936, *The Realm of the Nebulae* (New Haven: Yale Univ. Press).

Iye, M. 1997, *Proc. 7th Asian-Pacific Regional Meeting of IAU* (in press)

Iye, M. *et al.* 1996, *Proc. SPIE*, **2871**, (in press).

Jacoby, G.H. 1989, *Astrophys. J.*, **339**, 39.

Jacoby, G.H., Ciardullo, R., and Ford, H.C. 1990, *Astrophys. J.*, **356**, 332.

Jacoby, G.H., and Ciardullo, R. 1992, *Astrophys. J.*, **388**, 268.

Jacoby, G.H., Branch, D., Ciardullo, R., Davies, R.L., Harris, W.E., Pierce, M.J., Pritchet, C.J., Tonry, J.L., and Welch, D.L. 1992, *Publ. Astron. Soc. Pacific*, **104**, 599.

Jones, M. et al. 1993, *Nature*, **365**, 320.

Kaifu, N. 1996, *Proc. SPIE*, **2871**, (in press).

Kellermann, K.I. 1993, *Nature*, **361**, 134.

Kobayashi, S., Sasaki, S., and Suto, Y. 1996, *Publ. Astron. Soc. Japan*, **48**, L107.

Koo, D.C. *et al.* 1996, *Astrophys. J.*, **469**, 535.

Kraan-Korteweg, R.C., Cameron, L.M., and Tamann, G.A. 1988, *Astrophys. J.*, **331**, 620.

Kristian, J., Sandage, A., and Westphal, J. A. 1978, *Astrophys. J.*, **221**, 383.

Kundić, T. *et al.* 1997, submitted to *Astrophys. J.*.

Madore, B.F., and Freedman, W.L. 1991, *Publ. Astron. Soc. Pacific*, **103**, 933.

Mould, J. 1997, *Proc. 7th Asian-Pacific Regional Meeting of IAU* (in press)

Nakamura, T., and Suto, Y. 1995, *Astrophys. J.*, **447**, L65.

Narlikar, J.V. 1983, *Introduction to Cosmology* (Cambridge: Cambridge Univ. Press).

Nishimura, T. *et al.* 1996, *Proc SPIE*, **2871**, (in press).

Okamura, S. 1995, in *Scientific and Engineering Frontiers for 8-10m Telescopes*, eds. M. Iye and T. Nishimura (Tokyo: Universal Academy Press), p.33.

Pahre, M.A., and Mould, J.R. 1994, *Astrophys. J.*, **433**, 567.

Peacock, J.A., and Dodds, S.J. 1996, *Monthly Notices Roy. Astron. Soc.*, **280**, L9.

Phillips, M.M. 1993, *Astrophys. J.*, **413**, L105.

Pierce, M.J., Welch, D.L., McClure, R.D., van den Bergh, S., Racine, R., and Stetson, P.B. 1994, *Nature*, **371**, 385.

Riess, A.G., Press, W.H., and Kirshner, R.P. 1995, *Astrophys. J.*, **438**, L17.

Rowan-Robinson, M. 1985, *The Cosmological Distance Ladder* (New York: W.H. Freeman and Company)

Saha, A., Sandage, A., Labhardt, L., Tammann, G.A., Macchetto, F.D., and Panagia, N. 1996, *Astrophys. J.*, **466**, 55.

Sandage, A. 1988, *Ann. Rev. Astron. Astrophys.*, **26**, 561.

Sandage, A., Saha, A., Tammann, G.A., Labhardt, L., Schwengeler, H., Panagia, N., and Macchetto, F.D. 1994, *Astrophys. J.*, **423**, L13.

Sandage, A., Saha, A., Tammann, G.A., Labhardt, L., Panagia, N., and Macchetto, F.D. 1996, *Astrophys. J.*, **460**, L15.

Sato, K. 1981, *Monthly Notices Roy. Astron. Soc.*, **195**, 467.

Schaefer, B.E. 1996, *Astrophys. J.*, **460**, L19.

Schmidt, B.P., Press, W.H., and Kirshner, R.P. 1992, *Astrophys. J.*, **395**, 366.

Schmidt, B.P., Kirshner, R.P., Eastman, R.G., Phillips, M.M., Suntzeff, N.B., Hamuy, M., Maza, J., and Avilés, R. 1994, *Astrophys. J.*, **432**, 42.

Shectman, S., Landy, S.D., Demler, A., Tucker, D.L., Lin, H., Kirshner, R.P., and Schechter, P. 1996, *Astrophys. J.*, **470**, 172.

Smoot *et al.* 1991, *Astrophys. J.*, **371**, L1.

Strauss, M.A., and Willick, J.A. 1995, *Phys. Reports*, **261**, 271.

Sunyaev, R.A., and Zeldvich, Ya.B. 1972, *Comm. Astrophys. Space. Phys.*, **40**, 173.

Tammann, G.A., and Sandage, A. 1995, *Astrophys. J.*, **452**, 16.

Tanvir, N.R., Shanks, T., Ferguson, H.C., and Robinson, D.R.T. 1995, *Nature*, **377**, 27.

Tonry, J.L., and Schneider, D.P. 1988, *Astron. J.*, **96**, 807.

Tonry, J.L. 1991, *Astrophys. J.*, **373**, L1.

Tully, R.B., and Fisher, J.R. 1977, *Astron. Astrophys.*, **54**, 661.

van den Bergh, S., Abram, R.G., Ellis, R.S., Tanvir, N.R., Santiago, B.X., and Glazebrook, K.G. 1996, *Astron. J.*, **112**, 359.

Watanabe, M. 1996, *ph D thesis*, University of Tokyo.

Williams, R.E. *et al.* 1996, *Astron. J.*, **112**, 1335.

Yasauda, N., Fukugita, M., and Okamura, S. 1997, *Astrophys. J. Suppl.*, (in press)

Yoshii, Y., and Takahara, F. 1988, *Astrophys. J.*, **326**, 1.

# Quantum Cosmology:
# Problems for the 21st Century

James B. Hartle

*Institute for Theoretical Physics,*
*University of California, Santa Barbara, CA, 93108;*
*e-mail address: hartle@itp.ucsb.edu*

Two fundamental laws are needed for prediction in the universe: (1) a basic dynamical law and (2) a law for the cosmological initial condition. Quantum cosmology is the area of basic research concerned with the search for a theory of the initial cosmological state. The issues involved in this search are presented in the form of eight problems.

## 1   What are the Fundamental Laws?

Physics, like other sciences, is concerned with explaining and predicting the regularities of specific physical systems. Stars, the solar system, high-temperature superconductors, fluid flows, atoms, and nuclei are just some of the many examples. Beyond particular systems, however, physics aims at finding laws that predict the regularities exhibited universally by *all* physical systems — without exception, without qualification, and without approximation. These are the *fundamental* laws of physics. This essay is concerned with the fundamental law for the initial condition of the universe.

Ideas for the nature of the fundamental laws have varied as new realms of phenomena have been explored experimentally. However, until recently, all of the various ideas for fundamental laws have had one feature in common: They were proposals for dynamical laws — laws that predicted regularities in time. The laws of Newtonian mechanics, electrodynamics, general relativity, and quantum theory all have this character.

The Schrödinger equation is an example of fundamental dynamical law:

$$ i\hbar \, \frac{\partial \Psi}{\partial t} = H\Psi \ . \tag{1} $$

A fundamental theory of dynamics supplies the Hilbert space and the Hamiltonian operator $H$. However, a differential equation like (1) makes no predictions by itself. To solve (1), an initial condition — the state vector at one moment — must also be given. The Schrödinger equation then expresses the regularities in time that emerge from this initial state.

Where do the boundary conditions necessary to solve dynamical laws come from? In most of physics we study the evolution of subsystems of the universe and determine the boundary conditions by observation or experimental preparation. If we are interested in the evolution of the electromagnetic field in a room and observe no incoming radiation, we solve Maxwell's equations with no incoming radiation boundary conditions. To predict the probability for the decay of an atom prepared in an excited state, we solve the Schrödinger equation with that excited state as an initial condition at the time of preparation, and so on. Boundary conditions for the evolution of subsystems are obtained from *observations* of the rest of the universe outside the subsystem of interest.

Cosmology, however, presents us with an essentially different problem. The dynamical laws governing the evolution of the universe — the classical Einstein equation, for instance — require boundary conditions to yield solutions. But in cosmology, by definition, there is no "rest of the universe" to pass their specification off to. The cosmological boundary condition must be one of the fundamental laws of physics.

The inference is inescapable from the physics of the last sixty years that the fundamental laws are all quantum mechanical. If that is assumed, a theory of the initial condition is a theory of the universe's initial quantum state. The search for a fundamental theory of this initial cosmological quantum state is the aim of that area of basic research which has come to be called quantum cosmology.

A view thus emerges that there are two fundamental laws of physics:

- A theory of the basic dynamics,

- A theory of the initial condition of the universe.

Were the universe governed by the Schrödinger equation (1), the basic theory of dynamics would specify the Hamiltonian $H$; a theory of the initial condition would be a law for the initial quantum state.

The search for the fundamental dynamical law has been seriously under way since the time of Newton. Classical mechanics, Newtonian gravity, electrodynamics, special relativity, general relativity, quantum mechanics, quantum electrodynamics, quantum chromodynamics, electro-weak theory, grand unified theories, and superstring theories are but some of the important milestones in this search. By contrast, the search for a theory of the initial condition of the universe has been seriously under way for only a few decades. Why this difference? The answer lies in the empirical locality of the fundamental interactions on scales above the Planck length ($\sim 10^{-33}$cm), or put differently, the empirical fact that the fundamental interactions may be effectively described by a local quantum field theory on these scales. Assuming locality, the Hamiltonian of the whole universe can be deduced from experiments on familiar, laboratory, scales. However, typical ideas for the initial quantum state of the universe are non-local. They imply regularities *in space* that emerge mostly on large, cosmological scales. For example, the temperature of the cosmic microwave background is the same across the sky to one part in $10^5$, a distance which corresponded to $10^{20}$km at the time the radiation was emitted. It is only the recent progress in observational cosmology that has given us a picture of the universe on large enough scales of both space and time that is sufficiently detailed to confront with the predictions of a theory of the initial state of the universe.

## 2 Quantum Cosmology and The Everyday

Can those not interested in regularities on cosmological scales safely ignore the initial condition of the universe? Not if they seek a fundamental explanation of a number of its features we ordinarily take for granted. In this section we offer a few examples.

- **Isolated Subsystems**

    In one way we use a very weak theory of the initial condition every day. Many subsystems of the universe, in the laboratory and elsewhere, are approximately isolated for periods of time and can can be approximately described by solving the Schrödinger equation for the subsystem alone. In effect, we assume that for the purposes of making predictions about the subsystem, the wave function of the universe can be approximated by

$$\Psi(q^i, Q^A, t) \approx \psi(q^i, t)\Phi(Q^A, t) \tag{2}$$

where $q^i$ and $Q^A$ are coördinates referring to subsystem and the rest of the universe respectively and $\psi$ and $\Phi$ evolve *separately* under the Schrödinger equation. But what are the grounds for such an approximation? They do not lie in the nature of the Hamiltonian because that generally specifies interactions between all the coördinates. Rather the existence of isolated subsystems is a property of the quantum state. In discussing isolated subsystems, we are making weak quantum cosmological assumptions about the nature of this initial state.

- **The Quasiclassical Realm**

    Classical deterministic laws govern a wide range of phenomena in the universe over a broad span of time, place, and scale. This quasiclassical realm is one of the most immediate facts of our experience. But indeterminacy and distributed probabilities are the characteristics of a quantum mechanical universe. Classical deterministic dynamics can be but an approximation to the unitary evolution of the Schrödinger equation and the reduction of the state vector.

To what do we owe the validity of this approximation? In part it arises from a coarse-grained description with positions and momenta specified to accuracies well above the limitations of the uncertainty principle for instance. But coarse graining is not enough; there must also be some restriction on the initial state. Ehrenfest's theorem is a simple illustration of why. For the motion of a particle in one dimension, Ehrenfest's theorem relates the acceleration of the expected position to the expected value of the force:

$$ m \, \frac{d^2 \langle x \rangle}{dt^2} = - \left\langle \frac{\partial V(x)}{\partial x} \right\rangle \, . \tag{3}$$

This is generally true, but for *certain* states, typically narrow wave packets, the right hand side may be replaced by the force evaluated at the expected position to a good approximation resulting in the deterministic classical equation of motion

$$ m \, \frac{d^2 \langle x \rangle}{dt^2} \approx - \frac{\partial V(\langle x \rangle)}{\partial x} \, . \tag{4}$$

Just as only certain states lead to classical behavior in this simple model, so also only certain cosmological initial conditions will lead to the quasiclassical realm of familiar experience[1]. That too is a feature of the universe that must ultimately be traced to the initial condition.

• **Homogeneity of the Thermodynamic Arrow of Time.**

Isolated systems evolve towards equilibrium. That is a consequence of statistics. But in this universe presently isolated systems are mostly evolving towards equilibrium in the *same* direction of time. That is the homogeneity of the thermodynamic arrow of time. This is not a fact which can be explained by statistics or a property of the Hamiltonian alone for that is approximately time-reversal invariant. The homogeneity of the thermodynamic arrow of time follows from a fundamental law of the initial condition which mandates that the progenitors of today's isolated systems were all far from equilibrium in the early universe. As Boltzmann put it: "The second law of thermodynamics can be proved from the [time-reversible] mechanical theory if one assumes that the present state of the universe ... started to evolve from an improbable state"[2].

• **History.**

The reconstruction of history is useful for understanding the present in science as well as in human affairs. For example, we can best understand the character of biological species by understanding their evolution. We can best explain the present large scale distribution of galaxies by understanding how galaxies arose from tiny density fluctuations present shortly after the big bang. Such examples could be easily multiplied.

In physics, the reconstruction of history means using the fundamental laws to calculate the probabilities of alternative past events assuming the values of present records. Classically, present records *alone* are enough to calculate those probabilities by using them as the starting point for running the deterministic classical equations of motion backward in time. To reconstruct history in quantum mechanics, however, requires a theory of the initial condition *in addition* to present records.

The source of this difference between classical and quantum mechanics can be traced to the *arrow of time* in usual quantum theory.[a] Quantum mechanics treats the future differently from the past. To be sure, the Schrödinger equation (1) can be run backwards in time as well as forwards. But the Schrödinger equation is not the only law of evolution in quantum theory. In the usual story, when a measurement is made, the wave function is "reduced" by the action of the projection operator $P$ representing the outcome of the measurement, and then renormalized. This is a "second law of evolution":

$$\Psi \rightarrow \frac{P\Psi}{||P\Psi||} . \tag{5}$$

The evolution of the Schrödinger equation forwards in time is interrupted by (5) on a measurement. While the Schrödinger equation can be run backwards in time, the law (5) cannot, and that is a simple way of seeing the arrow of time in usual quantum mechanics. The same kind of arrow of time persists in more general quantum theories of closed systems where (5) is effectively used in in the construction of the probabilities of histories which are not necessarily of the outcomes of measurements.

How then does one calculate the probabilities of past events assuming present records in quantum mechanics? The simple answer is that one works forwards in time from the initial state. Evolving forwards using (1) and (5) one calculates the joint probabilities of *both* alternative events of interest in the past *and* the alternative values of the present records that follow them. From these one calculates the conditional probabilities of the past events given our particular present records in the usual way.

This process involves the initial state in an essential way. Strictly speaking, therefore, one cannot make any statements about the past without a theory of the universe's initial condition.

● **Phenomenology of the Initial Condition.**

While the above four everyday features of the universe are fundamentally traceable to the universe's initial quantum state, there is a large set of initial

---

[a]There are generalizations of quantum theory without an arrow of time in which the asymmetry of the usual theory may be understood as a difference between initial and final conditions[3]. We shall not consider these here.

184

states that would give rise to them. Put differently, the existence of isolated
subsystems together with the applicability of classical physics, the second law
of thermodynamics, and the possibility of historical explanation are not strong
constraints on the initial quantum state. Neither are the observations of large
scale features of the universe such as its approximate homogeneity and isotropy
or the fluctuations in the cosmic background radiation. The data are meager
and the Hilbert space of the observable universe is vast.

It would be possible to investigate quantum cosmology *phenomenologically*
by asking for the constraints present observations place on the initial state of
the universe. A density matrix $\rho$ is the way quantum mechanics represents the
statistical distribution of states with associated probabilities that would be
inferred. To investigate the initial condition phenomenologically is therefore
to ask for the density matrices consistent with observed features of the universe.

The observed features of the universe may not uniquely fix an initial condition but one should not exaggerate their weakness. The density matrix
$\rho = I/Tr(I)$, where $I$ is the unit matrix, is the unique representation of complete ignorance of the initial condition (*i.e.* no condition at all). But it also
corresponds to infinite temperature in equilibrium ($\rho \propto \exp(-H/kT)$) — an
initial condition whose implication of infinite temperature today is obviously
inconsistent with present observations.

The entropy $S/k = -Tr(\rho \log \rho)$ is a measure of the missing information
about the initial state in a density matrix $\rho$. Most of the entropy in the
matter in the visible universe is in the cosmic background radiation, a number
of order $S/k \approx 10^{80}$. As Penrose[4] has stressed, this is a large number, but
infinitesimally small compared to the maximum possible value of $S/k \approx 10^{120}$
if all that matter composed a black hole.

This essay, however, is not concerned with phenomenology. *Rather, it is
concerned with the fundamental law of the initial condition. We shall therefore
assume that the universe has a initial state $|\Psi\rangle$ and discuss the issues involved
in a search for the principles which determine it.*

## 3   Problems

Enumerating issues is one way of summarizing the present status of an area of
science, and motivating future research. Certainly, setting problems is a more
pleasant task than solving them, and quantum cosmology is such a young field
that it is easier to summarize problems than to survey accomplishments. It is
in this spirit that the author offers the following eight problems in quantum
cosmology:

## • Problem 1: What Principle Determines the Initial Condition of the Universe?

The evidence of the observations is that the universe was simpler earlier than it is now — more homogeneous, more isotropic, with matter more nearly in thermal equilibrium. This is evidence for a simple, discoverable initial condition of the universe. But what principles determine that initial state?

The most developed proposal for a principle determining the initial condition is the "no-boundary" wave function of Stephen Hawking and his associates[5]. The idea is that the initial condition of a closed universe is the cosmological analog of a ground state. This does not mean the lowest eigenstate of some Hamiltonian. Intuitively, the total energy of a closed universe is zero for there is no place outside from which to measure it. Correspondingly the Hamiltonian vanishes.

But the lowest eigenstate of a Hamiltonian is not the only way to find the ground state even in the elementary case of a particle moving in a potential $V(x)$. In that case, the ground state wave function may be expressed directly as a sum over Euclidean paths, $x(\tau)$:

$$\psi_0(y) = \sum_{\text{paths } x(\tau)} \exp\Big(-I[x(\tau)]/\hbar\Big) \qquad (6)$$

where $I = \int d\tau[m\dot{x}^2/2 + V(x)]$ is the Euclidean action. The sum is over paths $x(\tau)$ that have the argument of the wave function, $y$, as one end point, and a configuration of minimum action in the infinite past as another. Verify it for the harmonic oscillator for example!

This construction of a ground state wave function generalizes to closed universes. For definiteness suppose, for a moment, that the basic variables of the fundamental dynamical theory are the geometry of four-dimensional spacetime $\mathcal{G}$, represented by metrics on manifolds, together with matter fields such as the quark, lepton, gluon, and Higgs fields which we generically denote by $\phi(x)$. The arguments of cosmological wave functions are these basic variables restricted to spacelike surfaces, specifically the three-geometries of these surfaces, $^3\mathcal{G}$, and the field configurations on these surfaces, $\chi(\mathbf{x})$. The "no-boundary" wave function is of the form

$$\Psi_0[^3\mathcal{G}, \chi(\mathbf{x})] = \sum_{\mathcal{G}, \phi(x) \in C} \exp\Big(-I[\mathcal{G}, \phi(x)]/\hbar\Big) \qquad (7)$$

where $I[\mathcal{G}, \phi(\mathbf{x})]$ is the action for gravitation and matter. The "no boundary" wave function is specified by giving the class $C$ of four geometries $\mathcal{G}$, and matter fields $\phi(x)$ summed over in (7). So that the construction is analogous to (6), these geometries $\mathcal{G}$ should have Euclidean (signature $+ + ++$) and have one boundary at which they match the three-geometry where the wave function is evaluated. The matter fields must similarly match their boundary value. The defining requirement is that the $\mathcal{G}$'s have no *other* boundary, whence the name "no-boundary" proposal.

Nothing goes on in a typical ground state in a fixed background space-time. In field theory, the ground state is the time-translation invariant vacuum! However, this is not the context of the quantum cosmology of closed universes. Spacetime geometry is not fixed and there is therefore no notion of time-translation. Interesting histories therefore can happen; and the attractive nature of gravity makes things happen even in this cosmological analog of the ground state. In particular, initial, small, quantum, ground state fluctuations from homogeneity and isotropy that are predicted by this initial condition can grow by gravitational attraction to produce all the complexity in the universe that we see today.

This prescription for the "no-boundary" wave function is not complete. The reason is that the action $I[\mathcal{G}, \phi(x)]$ for gravitation coupled to matter is unbounded below. Were the sum in (7) extended over real, Euclidean geometries and fields, it would diverge! Rather, the sum must be taken over a class $C$ of *complex* geometries and fields. A complex contour of summation is, in fact, essential for the "no-boundary" wave function to predict the nearly classical behavior of geometry we observe in the present epoch. But many different complex convergent contours are possibly available and correspondingly there are many different "no-boundary" wave functions. These do not differ in their semi-classical predictions; but we still lack a complete principle for fixing this wave function of the universe.

The "no-boundary" idea has been described in terms of an effective theory of dynamics in which spacetime and matter fields are treated as fundamental variables. If spacetime is not fundamental, as in string theory or non-perturbative quantum gravity, then extending the idea to such theories becomes an important question. The essentially topological nature of the idea gives some hope that such an extension is possible.

The "no-boundary" wave function is not the only idea for a theory of the initial condition. Other notable candidates are the "spontaneous nucleation from nothing wave function"[6], and the ideas associated with the "eternally self-reproducing inflationary universe"[7]. Space does not permit a review of these and other theories, and the similarities and differences in their predictions. Discriminating between these and other ideas that may arise is certainly a problem for the 21st century.

• **Problem 2: How Can Quantum Gravity be Formulated for Cosmology?**

Gravity governs the evolution of the universe on the largest scales of space and time. That fact alone is enough to show that a quantum theory of gravity is required for a quantum theory of cosmology. Were the behavior of the universe on present cosmological scales all that was of interest, then a low energy approximation to quantum gravity would be adequate. Indeed most of the exploration of quantum cosmology has been carried out in such a low energy approximation assuming spacetime geometry and quantum fields are the basic variables with Einstein's theory coupled to matter as the basic action. Any divergences that arise are truncated in one way or another.

It is a reasonable expectation that low-energy, large scale, features of the universe, such as the galaxy-galaxy correlation function, are insensitive to the nature of quantum gravity on very small scales. But in quantum cosmology we aim not only at an explanation of such large scale features, but also at a theory of the initial condition adequate to describe the probabilistic details of the earliest moments of the universe. The inevitability of an initial singularity in classical Einstein cosmologies strongly suggests that the earliest moments of the universe will exhibit curvatures of spacetime characterized by the Planck length

$$\ell \equiv (\hbar G/c^3)^{1/2} \approx 10^{-33} \text{cm} \tag{8}$$

— the only combination of the three fundamental constants governing relativity, quantum mechanics, and gravity that has the dimensions of length. By making similar combinations with the right dimensions we can exhibit the Planck scales of energy and time. The universe at the epochs characterized by these scales will therefore depend on the detailed form of the fundamental quantum dynamical law for gravity.

There are a number of candidates for a finite, manageable quantum theory of gravity, notably superstring theory and non-perturbative canonical quantum gravity. However, neither of these theories is ready for application to quantum cosmology. String theory, for instance, exists in a practical sense as a set of rules for classical backgrounds and quantum perturbations away from them. Developing such theories to the point where they can be used for the non-perturbative quantum dynamics of closed cosmologies is thus an important problem.

The problem to be faced is not merely one of technique. Both of the approaches mentioned, and others as well, hint that spacetime geometry may not be a basic dynamical variable. If that is true, it becomes a conceptual issue just how to frame cosmological questions in the variables of the fundamental dynamical theory.

### • Problem 3: What is the Generalization of Quantum Mechanics Necessary for Quantum Gravity and Quantum Cosmology?

A generalization of usual quantum mechanics is needed for quantum gravity. That is because usual quantum mechanics relies in essential ways on a fixed, background spacetime geometry, in particular, to specify the notion of time that enters centrally into the formalism. This reliance on a fixed notion of time shows up in any of the various ways of formulating usual quantum theory — the idea of a state at a moment of time, the preferred role of time in the Schrödinger equation, the inner product at a moment of time, the reduction of the state vector at a moment of time, the commutation of fields at spacelike separated points, the equal time commutators of coördinates and momenta, *etc., etc.*

But in quantum gravity, spacetime geometry is not fixed, rather it is a quantum dynamical variable, fluctuating and generally without definite value. A generalization of usual quantum theory that does not require a fixed spacetime geometry, but to which the usual theory is a good approximation in situations when the geometry is approximately fixed, is therefore needed for quantum gravity and quantum cosmology. What, therefore, do we mean more generally by a quantum mechanical theory?

The most general objective of any quantum theory is the prediction of the probabilities of alternative, coarse-grained histories of the universe as a single, closed quantum mechanical system. For example, one might be interested in predicting the probabilities of the set of possible orbits of the earth around the sun. Any orbit is possible, but a Keplerian ellipse has overwhelming probability. Such histories are said to be coarse-grained because they do not specify the coördinates of every particle in the universe, but only those of the center of mass of the earth and sun, and these only crudely and not at every time.

However, the characteristic feature of a quantum mechanical theory is that consistent probabilities cannot be assigned to every set of alternative histories because of quantum mechanical interference. Nowhere is this more clearly illustrated than in the famous two-slit experiment shown in Figure 1. Electrons can proceed from an electron gun at left towards detection at a point $y$ on a screen along one of two possible histories — the history passing through the upper slit, $A$, and the history passing through the lower slit, $B$. In the usual story, probabilities cannot be assigned to these two histories if we have not *measured* which slit the electron passed through. It would be inconsistent to do so because the the probability to arrive at $y$ would not be the sum of the probabilities to arrive there going through the upper slit and lower slit:

$$p(y) \neq p_A(y) + p_B(y) \tag{9}$$

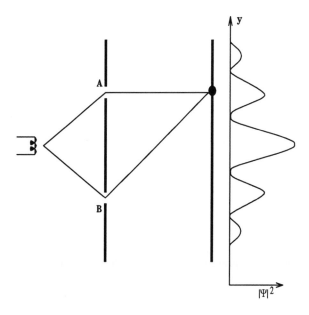

Figure 1: **The Two-Slit Experiment**

because of quantum mechanical interference. In quantum mechanics probabilities are squares of amplitudes and, of course,

$$|\psi_A(y) + \psi_B(y)|^2 \neq |\psi_A(y)|^2 + |\psi_B(y)|^2 \ . \tag{10}$$

A necessary consistency condition would not be satisfied.

A rule is thus needed in quantum theory to specify which sets of alternative histories may be assigned probabilities and which may not. In the usual, "Copenhagen" formulations of quantum mechanics presented in textbooks, probabilities can be assigned to the histories of alternatives of a subsystem that were "measured" by an "observer". But such formulations are not general enough for quantum cosmology which seeks to describe the early universe where there were neither measurements nor observers present.

In the more general quantum mechanics of closed systems[8] that rule is simple: probabilities can be assigned to just those sets of alternative histories for which there is vanishing interference between the individual members as a consequence of the system's initial state $|\Psi\rangle$. To make this quantitative we need the measure of this interference.

When there is a well-defined fixed notion of time, sequences of alternative sets of events at a series of times define a set of alternative histories. An individual history in such a set is a particular series of events, say $\alpha \equiv (\alpha_1, \alpha_2, \cdots, \alpha_n)$ at times $t_1 < t_2 < \cdots < t_n$. In usual quantum mechanics such a history is represented by a corresponding chain of (Heisenberg-picture) operators,

$$C_\alpha \equiv P_{\alpha_n}^n(t_n) \cdots P_{\alpha_2}^2(t_2) P_{\alpha_1}^1(t_1) , \tag{11}$$

time ordered from right to left. The application of the $C_\alpha$ to the initial state vector $|\Psi\rangle$ gives the branch state vector

$$C_\alpha |\Psi\rangle \tag{12}$$

corresponding to the history. Interference vanishes in a set of alternative histories when the branch state vectors corresponding to the different histories are mutually orthogonal. Sets of alternative histories with vanishing interference are said to *decohere*. The probabilities $p(\alpha)$ of the individual histories in a decoherent set are the squared lengths of the branch state vectors

$$p(\alpha) = ||C_\alpha|\Psi\rangle||^2 . \tag{13}$$

Decoherence insures the consistency of these probabilities.

Interference is thus measured by the *decoherence functional*:

$$D(\alpha', \alpha) = \left\langle \Psi | C_{\alpha'}^\dagger C_\alpha | \Psi \right\rangle \tag{14}$$

which becomes the central element in the theory. The condition of decoherence and the resulting probabilities may be expressed by the single formula

$$D(\alpha', \alpha) = \delta_{\alpha'\alpha} p(\alpha) . \tag{15}$$

The sets of possible coarse-grained histories, their decoherence functional, and (15) are the minimal elements of a quantum theory. A broad framework for quantum theories built on these elements, called generalized quantum mechanics, can be formulated in terms of decoherence functionals obeying general principles of Hermiticity, normalization, positivity, and the principle of superposition[9,10].

Histories represented by strings of projections at definite moments of time (11) and a decoherence functional (14) are the way that usual quantum theory implements the principles of generalized quantum theory. But there are many other ways, and among them are possibilities for generalizing usual quantum mechanics so that it works in the absence of a fixed spacetime geometry. Generalized sum-over-histories quantum theories have been discussed that put quantum theory into fully spacetime form with four-dimensional notions of histories, coarse grainings, and decoherence[10]. But the principles of generalized quantum mechanics are only a minimal set of requirements for quantum theory. What further principles determine the correct quantum mechanics for quantum gravity and quantum cosmology?

• **Problem 4: What are the Definite Predictions of the Initial Condition for the Universe on Large Scales?**

Extracting the predictions of a theory of the initial condition and comparing them with observations is a central problem in quantum cosmology. Predictions take the form of probabilities for present observations. The theory stands or falls on those predictions with probabilities sufficiently close to one (or zero) being observed (or not observed). These are called the *definite* predictions of the theory. We expect few of them. A simple, comprehensible, discoverable theory of the initial condition cannot predict all the complexity observed in the present universe with probability near one[11]. Rather, most predictions, as of the stock market, the weather, or the number of moons of Jupiter, will have more distributed probabilities based on the initial condition alone. (The vast majority will be nearly uniformly distributed which is no prediction at all.) In quantum cosmology one must search among the possible predictions for those which are predicted with probability near one. Interestingly, definite predictions may occur on all scales. For the purposes of simplicity we have divided the problem of what are the definite predictions of a theory of the initial condition into problems concerning regularities on cosmological, familiar, and microscopic scales.

Quantum cosmologists expect that a number of the general large scale features of the universe will be definite predictions of a theory of its initial condition. These include an approximately classical cosmological spacetime geometry after the Planck epoch, the approximate homogeneity and isotropy of the geometry and matter on scales above several hundred megaparsecs[b], the approximate spatial flatness of the universe (or what is the same thing its vast age in Planck units), the initial spectrum of quantum fluctuations which grew to become the galaxies, a sufficiently long inflationary epoch, and the cosmological abundances of the matter and radiation species.

The probabilities for these features of the universe arising from various theories of the initial condition have been explored in highly simplified models valid only in limited regions of the configuration space of possible present universes. The output of some of these calculations, such as the prediction of the spectrum of initial quantum fluctuations[12], are among the most success-

---

[b]The megaparsec is a convenient unit for cosmology. One megaparsec is 3.3 million light years. The size of the visible universe is several thousand megaparsecs.

ful achievements of quantum cosmology. But much more needs to be done to extend these calculations to the whole of configuration space with greater accuracy, generality and a precise quantum mechanical interpretation. That is a practical and immediate problem for quantum cosmology.

• **Problem 5: What are the Definite Predictions of the Initial Condition for Features of the Universe on Familiar Scales?**

We may treat this problem briefly because the obvious features of the universe on familiar scales that are traceable to the initial condition have been discussed qualitatively in Section II. However, those qualitative conclusions raise quantitative questions:

What are the coarse-grained variables defining a quasiclassical realm governed by deterministic laws and how are these variables related to the principle that determines the initial condition? How refined a quasiclassical description of the universe is possible before decoherence is lost or determinism is overwhelmed by quantum noise? How far in space and time can a quasiclassical description be extended? How do the phenomenological equations of motion that exhibit the determinism of the quasiclassical realm follow from the fundamental dynamical law, an initial condition, and an appropriate coarse-grained description? What is the connection of the coarse graining used to define a quasiclassical realm with that which is necessary to exhibit a second law of thermodynamics? How far out of equilibrium is the early universe in this coarse graining?

In short, a theory of the initial condition presents the challenge of defining *quantitatively* those features of the universe on familiar scales which are traceable, in part, to the nature of the initial condition.

• **Problem 6: What are the Definite Predictions of the Initial Condition on Microscopic Scales**

Our understanding of the world on microscopic scales above that set by the Planck length is summarized by the effective field theories which govern phenomena on these scales — for example, the standard model of elementary particle physics. However the forms of these effective field theories may be only distantly related to the form of the fundamental dynamical law. An analogous situation at a different scale may help explain why: The form of the Navier-Stokes equation which governs the dynamics of much of the quasiclassical realm is not easily guessed from the Lagrangian of the standard model of particle physics. In particular, the Navier-Stokes equation incorporates dissipation and depends on constitutive relations between density, pressure, temperature, viscosity, etc. — relations not contained in the Lagrangian of the standard model.

Of course, we understand qualitatively the relation between the laws of the standard model and the Navier-Stokes equation. The Navier-Stokes equation applies, not generally, not exactly, but only approximately in particular circumstances. It is *effective* equation with a limited range of approximate validity. In quantum mechanics particular circumstances are represented by the quantum state and a coarse-grained description. It is the quantum state whose special properties allow a classical approximation, set up the conditions for dissipation, determine the constituents, and allow for the local equilibrium from which the constitutive relations follow.

But the standard model itself may be only an effective approximation to a more fundamental dynamical law such as heterotic superstring theory or non-perturbative quantum gravity. We may therefore restate the problem of the definite predictions of the initial condition as follows: **What features of the effective dynamical laws that govern the elementary particle system at accessible energy scales are traceable to the cosmological initial condition and what to the fundamental dynamical law?** For instance, what is the origin of the locality of the effective interactions in a theory of the quantum state that is intrinsically non-local?

The investigations of the effects of wormholes by Hawking, Coleman, Giddings and Strominger, and others indicate just how strong the effect of the initial condition on the effective interactions could be[13]. Suppose that the sum over geometries defining the "no-boundary" wave function in (7) includes a sum over wormhole geometries — four dimensional geometries with many "handles" rather like a teacup has a handle. Suppose that the Planck scale (8) is the characteristic size of these wormholes in the geometries that contribute the most to the sum. Fields propagating in such geometries can go down a wormhole and emerge from one. On the much larger scales accessible to us, we would see the effect of Planck scale wormholes as local interactions which create and destroy particles. The net effect is to add to any local Lagrangian an infinite series of local interactions with coupling constants that are not fixed once and for all by the fundamental dynamical law or even by a renormalization procedure, but rather vary probabilistically with a distribution determined by the initial condition. If the distribution was sharp (as was hoped for the cosmological constant) then the couplings would be predicted.

A similar decoupling between the observed coupling constants and the basic Lagrangian would hold if the initial condition predicted domains of space much larger than our visible universe in which breaking of the symmetries of the fundamental dynamical law occurred in different ways in different places leading to a differing effective theories in different domains. The form of the effective theory governing our domain would then be only a probabilistic prediction of the fundamental dynamical law.

It has proved difficult to push such ideas very far, but their lesson is clear. The form and couplings of the effective interactions at accessible scales may be probabilistically distributed in a way which depends on the initial condition. Finding these distributions and how sharp they are is therefore an important problem in quantum cosmology.

- **Problem 7: What Does Quantum Cosmology Predict for IGUSes?**

Most of predictions of the initial condition that we have considered so far are described in terms of alternatives of the quasiclassical realm. But there are many sets of decohering histories of the universe arising from a theory of its initial condition and dynamics that have nothing to do with the usual quasiclassical realm. These sets may be quantum mechanically incompatible with each other and with the usual quasiclassical realm in the sense that pairs of them cannot be combined into a common decohering set. Such incompatible sets are not contradictory; rather they are complementary ways of viewing the unfolding of the initial condition into alternative histories. The quantum mechanics of closed systems does not distinguish between such incompatible sets of alternative histories except by properties such as their classicality. All are in principle available for the process of prediction.

Yet, as observers, we describe the universe almost exclusively in terms of the familiar variables of classical physics. What is the reason for this narrow focus in the face of all the other non-quasiclassical decohering sets of alternative histories? Some see this disparity between the possibilities allowed by quantum theory and the possibilities utilized by us as grounds for augmenting quantum mechanics by a further fundamental principle that would single out one decohering set of histories from all others[14]. That is an interesting line of thought, but another is to seek an explanation within the existing quantum mechanics of closed systems.

Human beings, bacteria, and certain computers, are examples of information gathering and utilizing systems (IGUSes). Roughly, an IGUS is a subsystem of the universe that makes observations and thus acquires information, makes predictions on the basis of that information using some approximation (typically very crude) to the quantum mechanical laws of nature, and exhibits behavior based on these predictions[15]. To explain why IGUSes are exhibited by the universe, or why they behave the way they do, or to answer questions like "Why do we utilize quasiclassical variables?", one must seek to understand how IGUSes evolved as physical systems within the universe. In quantum cosmology that means examining the probabilities of a set of histories that define alternative evolutionary tracks. For IGUSes that can be characterized in terms of alternatives of the usual quasiclassical realm, it is a plausible conjecture that they evolved to focus on the usual quasiclassical alternatives because these present enough regularity over time to permit prediction by relatively simple models (schemata). This would be one kind of explanation of why we utilize the usual quasiclassical realm. However, we should not pretend that we are close to being able to carry out a calculation of the relevant probabilities or even likely to be in the early 21st century!

But what of sets of histories that are completely unrelated to the usual quasiclassical realm? Might some of these exhibit IGUSes with high probability that make predictions in terms of variables very different from the familiar quasiclassical ones? Or is the usual quasiclassical realm somehow distinguished with respect to exhibiting IGUSes? To answer such questions one would need a general characterization of IGUSes that is applicable to all kinds of histories — not just quasiclassical ones — and an ability to calculate the probabilities of various courses of the IGUSes' evolution. Such questions, while quite beyond our power to answer in the present, illustrate the range of predictions in principle possible in a quantum universe from a fundamental theory of dynamics and the initial condition of the universe.

## 4 Unification

The universal laws that govern the regularities of every physical system are one goal of physics. A fundamental dynamical law is one objective. Quantum cosmology is concerned with the equally necessary fundamental law specifying the initial condition of the universe.

Historically, many of the advances towards the fundamental laws have had in common that some idea that was previously thought to be universal was subsequently seen to be only a feature of our special place in the universe and the limited range of our experience. With more data, the idea was seen to be a true physical fact, but one which is a special situation in a yet more general theory. The idea was a kind of "excess baggage" which had to be jettisoned to reach a more general, more comprehensive, and more fundamental perspective[16].

It is not difficult to cite examples of such excess theoretical baggage in the history of physics: the idea that the earth was the center of the universe, the idea of Newtonian absolute time, the idea that the increase entropy was a basic dynamical law, the idea that spacetime geometry is fixed, the idea of a classical world separate from quantum mechanics, *etc. etc.* Further, and more importantly for the present discussion, one can cite examples concerning the nature of the fundamental laws themselves: the idea that thermodynamics was separate from mechanics, the idea that electricity was separate from magnetism, and more recently the idea that there were separate weak and electromagnetic interactions. These seemingly distinct theories were eventually unified. Today, extrapolations of the standard model of the electro-weak and strong interactions suggest a unified theory of these forces characterized by an energy scale a little below the Planck scale.

Examples such as those just cited have led some physicists to speculate that the existing separation between the dynamical laws for the gravitational and other forces is also an example of excess baggage arising from the limitations of present experiments to energies well below the Planck scale, and to search for a unified fundamental law for dynamics of *all* the forces. Secure in the faith that fundamental laws are mathematically simple, heterotic superstring theory or its extensions have been the inspiring results.

However, such a unified *dynamical* law does not really deserve the common designation of "a theory of everything" or a "final theory". Quantum cosmology offers a further opportunity for unification beyond dynamical laws. Could it be that the apparent division of the fundamental laws into a law for dynamics and a law for an initial quantum state is a kind of excess baggage similar to those described above? Gell-Mann[17] has stressed that there is already an element of unification in ideas such as the "no-boundary" proposal. In (7) the same action that determines fundamental dynamics also determines the quantum state of the universe. Despite this connection, the "no-boundary" proposal is a separate principle specifying one wave function out of many possible ones. Thus we have an eighth problem for quantum cosmology:

• **Problem 8: Is there a Fundamental Principle that would Single Out *Both* a Unified Dynamical Law *and* a Unique Initial Quantum State for the Universe? Could that same Principle Single Out the Form of Quantum Mechanics from Among Those Presented by Generalized Quantum Theory?**

In such a unification of the law of dynamics, the cosmological boundary condition, and the principles of quantum mechanics, we would, at last, have a truly unified fundamental law of physics governing the universe as a whole and everything within it. That is truly a worthy problem for physics in the twenty-first century!

## 5 Further Reading

An article introducing quantum cosmology appears in *Scientific American*[18]. An accessible but more advanced introductory review is included in the 1989 Proceedings of the Jerusalem Winter School[19]. That article contains a nearly exhaustive list of references at the time and a guide to the literature. For an introduction to the quantum mechanics of closed systems, see the work by the author[8]. For an exposition of the applications of quantum mechanics to cosmology see an article by the author in the Proceedings of the 1992 Les Houches Summer School[10].

## Acknowledgments

Thanks are due to G. Horowitz and J. West for critical readings of the manuscript. Preparation of this essay was supported in part by the US National Science Foundation under grants PHY95-07065 and PHY94-07194.

## References

[] No attempt has been made to assemble a complete list of references to the topics touched on in this essay. Rather, the emphasis has been placed on reviews of topics that are most useful directly to the points considered. These are not always the earliest nor the latest references.

1. For a more quantitative discussion see, *e.g.* J.B. Hartle, *Quasiclassical Domains In A Quantum Universe*, in *Proceedings of the Cornelius Lanczos International Centenary Conference*, North Carolina State University, December 1992, ed. by J.D. Brown, M.T. Chu, D.C. Ellison, R.J. Plemmons, SIAM, Philadelphia, (1994); LANL e-print gr-qc/9404017.

2. L. Boltzmann, *Zu Hrn. Zermelo's Abhandlung Über die mechanische Erklärung irreversibler Vorgange*, Ann. Physik, **60**, 392, (1897).

3. M. Gell-Mann and J.B. Hartle, *Time Symmetry and Asymmetry in Quantum Mechanics and Quantum Cosmology*, in *Proceedings of the NATO Workshop on the Physical Origins of Time Asymmetry, Mazagón, Spain, September 30-October 4, 1991* ed. by J. Halliwell, J. Pérez-Mercader, and W. Zurek, Cambridge University Press, Cambridge (1993); LANL e-print gr-qc/9309012.

4. R. Penrose, *Singularities and Time Asymmetry* in *General Relativity: An Einstein Centenary Survey* ed. by S.W. Hawking and W. Israel, Cambridge University Press, Cambridge (1979).

5. S.W. Hawking, *The Quantum State of the Universe*, Nucl. Phys. B, **239**, 257 (1984).

6. A. Vilenkin, *Predictions from Quantum Cosmology*, in *String Gravity and Physics at the Planck Scale*, ed. by N. Sanchez and A. Zichichi (Kluwer Academic, Dordrecht, 1996) and *Approaches to Quantum Cosmology*, Phys. Rev. D50, 2581 (1994).

7. A.D. Linde, *Particle Physics and Inflationary Cosmology*, (Harwood Academic Publishers, Chur, Switzerland 1990); A.D. Linde, D.A. Linde, and

A. Mezhlumian, *From the Big Bang Theory to the Theory of a Stationary Universe*, Phys. Rev. D **49**, 1783 (1994).

8. For expositions see, *e.g.* J.B. Hartle, *The Quantum Mechanics of Closed Systems*, in *Directions in General Relativity, Volume 1: A Symposium and Collection of Essays in honor of Professor Charles W. Misner's 60th Birthday*, ed. by B.-L. Hu, M.P. Ryan, and C.V. Vishveshwara, Cambridge University Press, Cambridge (1993); LANL e-print gr-qc/9210006; and R. Omnès, *The Interpretation of Quantum Mechanics*, (Princeton University Press, Princeton, 1994).

9. C.J. Isham, *Quantum Logic and the Histories Approach to Quantum Theory. J. Math. Phys.*, **35**, 2157 (1994).

10. J.B. Hartle, *Spacetime Quantum Mechanics and the Quantum Mechanics of Spacetime* in *Gravitation and Quantizations*, Proceedings of the 1992 Les Houches Summer School, edited by B. Julia and J. Zinn-Justin, Les Houches Summer School Proceedings, Vol. LVII, North Holland, Amsterdam (1995); LANL e-print gr-qc/9304006.

11. J.B. Hartle, *Scientific Knowledge from the Perspectives of Quantum Cosmology* in *Boundaries and Barriers : On the Limits to Scientific Knowledge*, edited by John L. Casti and Anders Karlqvist, Addison-Wesley, Reading, Mass., 1996; LANL e-print gr-qc/9601046.

12. J. Halliwell and S.W. Hawking, *Origin of Structure in the Universe*, Phys. Rev. D, 31, 1777, 1985.

13. For a review see, A. Strominger, *Baby Universes*, in *Quantum Cosmology and Baby Universes: Proceedings of the 1989 Jerusalem Winter School for Theoretical Physics*, eds. S. Coleman, J.B. Hartle, T. Piran, and S. Weinberg, World Scientific, Singapore (1991).

14. H.F. Dowker and A. Kent, *On the Consistent Histories Approach to Quantum Mechanics, J. Stat. Phys.* 82, 1574, (1996), LANL e-print gr-qc/9412067.

15. IGUSes are complex adaptive systems in the context of quantum mechanics. For more on the general characterization of complex adaptive systems see, M. Gell-Mann, *The Quark and the Jaguar*, W. Freeman San Francisco (1994).

16. J.B. Hartle, *Excess Baggage*, in *Elementary Particles and the Universe: Essays in Honor of Murray Gell-Mann* ed. by J. Schwarz, Cambridge University Press, Cambridge (1990).

17. M. Gell-Mann, *Dick Feynman — the Guy in the Office Down the Hall*, *Physics Today*, **42**, no. 2, 50, (1989).

18. J. Halliwell, *Quantum Cosmology and the Creation of the Universe*, *Scientific American*, **265**, no. 6, 76, (1991).

19. J. Halliwell, *Introductory Lectures on Quantum Cosmology* in *Quantum Cosmology and Baby Universes: Proceedings of the 1989 Jerusalem Winter School for Theoretical Physics*, eds. S. Coleman, J.B. Hartle, T. Piran, and S. Weinberg, World Scientific, Singapore (1991) pp. 65-157.